Aircraft & Rocket Turbines

Advances in Physics and Design

by

A.H. Epstein, Ph.D.
E.M. Greitzer, Ph.D.
J.L. Kerrebrock, Ph.D.
J. Guenette, Ph.D.
J. Paduano, Ph.D.
C.S. Tan, Ph.D.

Gas Turbine Laboratory
Department of Aeronautics and Astronautics
Massachusetts Institute of Technology

J. Vennard
Publication Editor

Wexford
2008

TABLE OF CONTENTS

1 SUMMARY

This book is comprised of research in five technical areas of aircraft and rocket turbine design:

(1) Flutter clearance in aircraft engine compressors

(2) Physics of coolant/main flow interactions in advanced aircraft and rocket turbines

(3) Tip clearance flow effects in axial and centrifugal turbomachines

(4) Use of an active rotor to determine parametric dependence of aerodynamic damping

(5) Fabrication and Testing of a Spur-Actuated Active Compressor Rotor Blade

A common theme threaded through these research areas is three dimensional and/or unsteady flow phenomena, and the enhancement of overall performance of turbomachinery through flow management and design. In other words, the primary challenge to achieving increased overall performance of turbomachines is linked to local phenomena which are inherently three-dimensional and/or unsteady.

1.2 Research Abstracts

1.2.1 Flutter Clearance in aircraft engine compressors

A framework for flutter operability assessment, based upon a new set of similarity parameters, has been developed. This set consists of four parameters which embrace both the performance characteristics in terms of corrected mass flow and corrected speed, and the flight condition in terms of inlet temperature and density (or, equivalently, inlet pressure). It is shown that a combined mass-damping parameter, g/ρ^*, novel in the field of turbo machinery aeroelasticity, can summarize the individual effects of mechanical damping, g, and blade mass ratio, $\mu(\propto 1/\rho^*)$. A particular selection of four non-dimensional parameters, including g/ρ^* and a compressible reduced frequency parameter, K^*, allows for a decoupling of corrected performance effects from purely aeroelastic effects, for a given machine and a specific mode shape. This view of flutter operability is applied to the analysis of a set of full-scale engine data. The data exhibits the trend that increasing K^* and increasing g/ρ^* have stabilizing effects, which is consistent with previous work on flutter stability. We propose that these trends hold generally, and apply the trends towards constructing a flutter clearance methodology, a test procedure that satisfies the requirements for comprehensive flutter stability testing.

1.2.2 Physics of coolant/main flow interactions in advanced aircraft/rocket turbines

Modern high performance gas turbine engines utilize film cooling to reduce the heat load on high-pressure turbine stage components, thereby increasing the maximum turbine inlet temperature at which the cycle can operate. The aerodynamic performance of a film cooled turbine stage and the loss caused by film cooling have been measured and quantified in a series of test carried out in the MIT Blow-down Turbine Facility. An un-cooled turbine stage was first fabricated with solid blading and tested using a newly developed short duration measurement technique. The stage was then modified to incorporate vane, blade and rotor casing film cooling. The film-cooled stage was then tested over a range of coolant-to-mainstream mass flow and temperature ratios for the same range of operating conditions (pressure ratios and corrected speeds) as the un-cooled turbine.

1.2.3 Role of Tip Clearance Flow on Axial Compressor Stability

An examination of the fluid dynamic phenomena that link tip clearance flow to the formation of short length-scale (spike) rotating stall disturbances has been carried out. It is found that the onset of growth in tip clearance blockage characterizes the lowest flow coefficient for which a steady blade passage solution exists. It is also found that this condition leads to the formation of spike disturbances. A scenario and criteria for this tip clearance blockage behavior are proposed based on trailing edge backflow and leading edge spillage to the adjacent blade passage. Both are associated with tip clearance flow and occur below the blade tip. Trailing edge backflow involves tip clearance fluid from adjacent blade passages. The leading edge spillage consists of tip clearance fluid from the local blade passage. These two criteria explain the observed length-scale of spike disturbances. This scenario is consistent with several experimental observations on axial compressor stall inception. The implications of these results on the role of single blade passage computations in stall prediction and on the effectiveness of techniques used to delay stall are also discussed.

1.2.4 Impact of Tip Clearance Flow on Centrifugal Impeller Pump Performance

A computational study has been conducted to evaluate the effects of tip-clearance on centrifugal pump impeller efficiency and pressure rise and to identify specific phenomena that lead to performance decrease with increased tip-clearance. Analysis of the flow field shows that there are two principal mechanisms responsible for the performance changes with tip-clearance. First, tip-leakage flow partially mixes with the core flow when it enters the channel. Second the resulting low streamwise velocity region (compared to the core flow) is responsible for flow blockage and decreased pressure rise. The tip-vortex structure associated with the tip-clearance flow has also been examined. The tip-vortex, which dominates the secondary flow patterns, increases in size with downstream distance. Low streamwise velocity flow is embedded in the tip vortex flow region. A simple analysis of the tip-vortex center trajectory has been developed which enables prediction of the wake flow trajectory. A model for the tip-clearance related decrease in pressure rise and efficiency has been developed, which accounts for both mixing and blockage effects. The model has been assessed against CFD results as well as published results and is found to capture performance sensitivity to tip-clearance variation for the examined data. The model also provides approximate analytical expressions for efficiency sensitivity.

1.2.5 Fabrication and Testing of a Spar-Actuated Active Compressor Rotor Blade

Design and development of rotor blades for an active aeroelastic rotor has been an ongoing research activity in the Gas Turbine Laboratory. New prototypes of an active rotor blade design were tested in the MIT/GTL spin test facility during 2002 and 2003. These blades achieve an industrial transonic blade shape (GE "Fan C") using graphite-epoxy spars and a high strength Rohacell core. Active control is achieved through piezo actuation and strain gauge sensing. The active rotor concept provides a vehicle for demonstration of modeling, passive control, and active control of aeroelastic phenomena in turbomachinery. Significant testing and development of this concept were completed under the program. Active rotor research has concentrated on the technologies and processes that would make an active rotor feasible, and on quantitative determination of the bending and twisting displacement capability of the device. Results are outlined in greater detail in Section 7.

2 *INTRODUCTION*

A framework for flutter stability was developed for the case of an axial-flow machine with specified flowpath geometry and a particular modeshape. Such a framework can be described in terms of four nondimensional parameters, by combining the mechanical damping and the fluid inertia into a single reduced damping parameter, g/ρ^*, a novel development in the context of turbo machinery flutter. The framework can be applied to the requirement of flutter clearance via testing.

Newly developed short duration measurement techniques have enabled cost-effective studies on influence of film cooling on the aerodynamic performance of a turbine stage. The performance of a un-cooled stage was measured and the influence of the rotor tip gap quantified. The stage efficiency was reduced by 1.7% for each percent increase in rotor tip gap. For the film-cooled stage, the stage efficiency was reduced by 1.75% (at the reference coolant mass flow and temperature ratio) when compared to the baseline. Each additional percent of cooling flow reduced the stage efficiency by 0.5%.

The connection of short length-scale stall inception to two specific simultaneous threshold flow events on axial compressor blade passage level has been established. These two threshold flow events are trailing edge backflow that involves tip clearance fluid from adjacent blade passages and the leading edge flow spillage that consists of tip clearance fluid from the local blade passage. These two events explain the observed length-scale of spike disturbances (~ two to three blade pitches). The results also suggest the utility of single-blade passage computations for stall inception prediction and effective means of delaying stall.

In centrifugal pump the secondary flow structure in an unshrouded impeller is dominated by tip vortex. The two principal mechanisms responsible for the performance changes with tip-clearance are: (1) tip-leakage flow partially mixes with the core flow when it enters the channel; and (2) the resulting low streamwise velocity region (compared to the core flow) is responsible for flow blockage and decreased pressure rise. A model for the tip-clearance related decrease in pressure rise and efficiency has been developed, which accounts for both mixing and blockage effects. It is deduced from the model that impeller efficiency sensitivity increases with increasing blade exit angle, decreasing flow coefficient and decreasing number of blades.

Finally, active rotor concept testing was completed during the Spring of 2003. The technologies and processes needed for an active rotor have been demonstrated with the completion of a composite rotor blade with the strength and actuation properties required for aeroelastic diagnostics, system identification, mistuning studies, and control. The effectiveness of this rotor was quantified in terms of bending and twisting displacement, both in bench tests and in situ (rotating environment). Results indicate that for broadband excitation of aeroelastic modes associated with the rotor first bending mode (that is, for system identification of the associated blade row aerodynamics), the active rotor achieves about half the desired excitation level of 0.5 mm. For the first twist mode, the active rotor achieves less than half the desired twist of 0.5 degrees. On the other hand, for demonstration of active damping of rotating stall, or mistuning, an active rotor would be more than adequate, since these types of tests would be conducted at or near the natural frequency of the active rotor blades, where displacements are large. Thus the way forward to a full active rotor facility has been developed.

3 COMPRESSOR FLUTTER CLEARANCE METHODOLOGY

(J. L. Kerrebrock)

3.1 Introduction

To assure reliability and safety of jet propulsion, the potential for blade flutter must be eliminated from the turbomachinery stages. Rig and engine testing are therefore necessary in any engine development program to ensure that flutter does not occur in the operational regime. The relevant regime for assessing performance (i.e., pressure ratio and efficiency) is typically measured in terms of corrected mass flow, m_c , and corrected speed. For a given machine, these are equivalent to the axial and tangential blade relative Mach number, respectively. Flutter stability, however, is not solely described by these variables, but requires other variables as well. In particular, the flight condition in terms of inlet temperature and pressure is known to influence flutter stability. This task addressed the effects upon flutter stability from the full (non-dimensional) set of parameters which span the operating space for a given flow-path geometry.

The problem of flight condition effects upon flutter stability was first identified by Jeffers and Meece [1975] in the context of a fan flutter problem during the development of the F100 engine. While flutter tests at sea level ambient conditions initially suggested that the engine would not flutter; subsequent flight and ground tests confirmed that changes in inlet temperature and density led to a flutter problem in the initial design. Further laboratory testing [Mahalic et al. 1977] corroborated these results on the F100. Such behavior is shown schematically in Figure 3.1. A more comprehensive parameter study of the effects of flight condition was included in experiments on an annular cascade [Jutras et al. 1980] which focused upon choke flutter in mid-stage compressors. That study also showed temperature effects to be a significant contributor to stability, but did not find inlet density to be a significant factor. Using the results derived in this Task, the discrepancy in the density effects between these studies can be attributed to differing blade root attachments (as will be explained in the following). Stargardter [1979] showed that the classical procedure of correlating flutter onset with reduced velocity, U^*, and flow incidence angle was not valid in his tests. Since Stargardter's tests were conducted at constant inlet temperature and density, increases in reduced velocity, U^*, resulted in Mach number increases as well. The classical correlation procedure could not account for such mixing of Mach number effects and reduced velocity effects, which underscores the need to account for *all* the relevant parameters in assessing aeroengine flutter stability.

Task 1 focused upon developing a rational methodology for flutter clearance; that is, for determining that flutter does not occur in the intended operating regime of a given engine. To accomplish this, we develop a minimal parameter space in terms of similarity parameters. Further, parameters are selected which separate the effects of performance point and of flight condition. A set of full-scale engine test data, provided by the Volvo Corporation, is analyzed using this new framework. Finally, we describe a flutter clearance procedure.

3.2 Stability Boundary Representation

For flutter clearance, we must assess the flutter stability boundary of a given machine. The six parameters, Mach number, M, flow angle, α, reduced frequency, k, temperature, T, density, ρ, and blade mechanical damping, g are needed to characterize the flutter boundary. We propose that a set of four similarity parameters that are especially useful for stability boundaries. These four are corrected mass flow, m_c , corrected speed, N_c , and two new parameters, reduced damping, g/ρ^*, and compressible reduced frequency, K^*. Note that the vibrational mode-shapes are presumed to be specified, and that each relevant mode-shape should be considered.

3.2.1 Development of Parameters

The parameter development begins with the linear stability criterion for single-mode flutter of a tuned rotor. Note that complications such as mistuning (with a specified amount and pattern) and nonlinearities will not change the governing parameters, though they can alter the form of the stability criterion. We begin with the structural dynamic equation (Crawley [1988], or Főrsching [1994]) of a specific mode, η, on blade j

$$\frac{1}{\omega_0^2}\ddot{\eta}_j + (1 + ig)\eta_j = \frac{F_j}{cm_0\omega_0^2} \tag{3.1}$$

where ω_0 is the modal natural frequency, F is the fluid force, c is the chord, and m_0 is the modal mass. The imaginary damping term, ig, presumes that η is complex with an imaginary part phase-lagged by 90° from the real part (true for positive frequencies).

For a tuned rotor, the use of interblade phase coordinates (Lane, 1956) decouples the fluid forces between blades. For interblade phase, σ_j , we use the following model for the fluid forces

$$\frac{F_{\sigma_j}}{c^2\rho U^2} = [l_{\sigma_j}(M, \alpha, k)]\eta_{\sigma_j} + h.o.t \tag{3.2}$$

where U is inlet blade-relative velocity, k is the reduced frequency, and l is a non-dimensional force coefficient. For the linear stability criterion, the higher order terms are neglected.

Substituting the expression (3.2) into the interblade-phase version of the structural equation, (3.1), one obtains a linear, second-order equation for η. Using a harmonic time-dependence, $\eta=\eta_0 e^{i\omega t}$, one obtains the following equation for ω,

$$-\left(\frac{\omega}{\omega_0}\right)^2 \frac{1}{\rho}$$

where ρ^* is a fluid inertia parameter, $c^3 \rho/m_0$, which is equivalent to the traditionally defined mass ratio, $\mu = 4/\pi\rho^*$. We prefer ρ^* since it highlights the dependence upon inlet density. For linear stability, Im(ω), must be positive for all possible interblade-phases, σ_j . Taking the imaginary part of (3.3), we find the following criterion for stability:

$$g / \rho^* > \frac{1}{k^2} \max_{\sigma_j} \left\{ \mathrm{Im}\, l_{\sigma_j}(M, \alpha, k) \right\} \tag{3.4}$$

This demonstrates that the parameter space is four-dimensional: (g/ρ^*, k, M, α), and that the separate parameters of damping, g, and fluid inertia, ρ^*, may be *combined into a single parameter* for purposes of stability. We term this new parameter, g/ρ^*, the reduced damping. Although this is novel in the context of turbomachinery flutter, it is not an entirely new idea, and has been applied in bluff-body interactions (Vickery and Watkins, 1964, and Scruton, 1965). It should be recognized that the vibration-induced fluid force coefficient, $l_{\sigma j}$, is independent of g/ρ^*, but still depends on M, α, and k.

3.2.2 Implications of g/ρ^*

This result has some important implications to aeroengine flutter. For example, the reduced damping parameter summarizes the interdependent effects of mass and damping as discussed in the parameter study of Forsching (1994).

An estimate of the fluid inertia at sea level is $\rho^* \sim O(0.01)$, for a metal fan blade in bending. The mechanical damping depends upon whether the vibrating blade rubs against other surfaces, causing friction. Srinivasan (1981, 1987) estimates that the frictional damping in bending for inserted metal blades is of order $g \sim O(0.01)$, while material damping alone (i.e. no friction) gives $g \sim O(0.0001)$. Thus, with friction, g/ρ^* is of the same order as the fluid term (3.3) and is significant towards assessing stability. Without friction, g/ρ^* is too small and therefore is not a significant factor.

This observation helps explain the discrepancy in density effects between the experiments on the F100 and the annular cascade experiments, as mentioned in the introduction. The F100 engine had the inserted blades with frictional damping, while the annular cascade had a special root attachment (described in Rukowski et. al., 1978) with no rubbing or friction. Thus, the presence of density effects upon flutter stability in the F100, should be expected since $g/\rho^* \sim O(1)$, while the absence of density effects in the annular cascade should also be expected since $g/\rho^* \sim O(0.01)$ was too small to be relevant in this case. Note that these "density effects" (i.e. g/ρ^* effects) are considered for constant M, α and k.

3.2.3 Decoupling performance from flight condition

Using the parameter space developed above, we can fully describe flutter stability for a given structural mode and flowpath geometry, throughout its operational regime. Although the above parameters (M, α, k, and g/ρ^*) span this space, we find it more convenient to use another set of four parameters for operability assessment. First, it is useful to use the corrected mass flow, m_c and corrected speed, N_c, which are equivalent to Mach number, M, and flow angle, α, for a given geometry.

The main reason, however, is to separate the performance effects from the flight condition effects. Specifying the modal parameters (which sets ω_0) and the flight condition (which fixes T), the parameters M and k become linearly proportional, making their individual effects indistinguishable. To address this, we replace k with K, a parameter which depends only upon the modal parameters and the flight condition, defined as

$$K^* = \frac{k}{M} = \frac{\omega_0 c}{\sqrt{\gamma RT}} \tag{3.5}$$

where γ is the ratio of speci£c heats, R is the gas constant, and T is the temperature. Physically, K is the ratio between the acoustic timescale and the oscillation timescale. For constant modal parameters (m_0,g,ω_0), using the parameter set (m_c,N_c,K^*,g/ρ^*) decouples the performance effects from the flight condition. The first two parameters, (m_c, N_c), alone account for the corrected performance, and the latter two, (K^*,g/ρ^*), alone account for the flight condition in terms of inlet temperature and density. In the case of a front stage with an ideal inlet and given modal parameters, T and ρ can be expressed in terms of the aircraft flight Mach number, M_f, and altitude, a. Figure 3.2 shows this relationship for a typical supersonic aircraft (taken from McCormick, 1995).

Using K^*, the stability criterion (3.4) can be rewritten as follows:

$$g / \rho^* > \frac{1}{K^{*2}} \max_{\sigma_j} \left\{ l'_{\sigma_j} (M, \alpha, K^*) \right\} \tag{3.6}$$

where l is the fluid force, F, non-dimensionalized by $\gamma c^2 \ p_0$, with p_0 being the inlet pressure. It can be shown (see Khalak,2000, for details) that for constant modal parameters, K^*, or a function thereof, is the only replacement for k which decouples the performance from the flight condition and satisfies a stability criterion of the form (3.6).

3.2.4 Varying Modal Parameters

Sometimes the modal parameters vary with operating condition, notably in the case of centrifugal stiffening. To maintain the separation of performance effects and flight condition effects, K^* and g/ρ^*, which were defined for constant modal parameters, should be generalized. In the case of centrifugal stiffening, for example, we can generalize K^* using the *non-rotating* natural frequency (i.e. at rest) in the definition of K^*. This keeps only the speed-independent effects of frequency in K^*, and maintains the separation of performance from flight condition. Generalizations to other dependencies of the modal parameters should be based on the specific form of the dependency.

3.2.5 Flutter boundaries

The four parameters, (m_c, N_c, K^*, g/ρ^*), offer a view of flutter stability which completes the performance map view of Figure 3.1, in which the flutter boundary moves with changes in inlet temperature. In general, the flutter boundary on the performance map can move with changes in K^* and g/ρ^*. A useful diagram is a simultaneous view of the performance map, of pressure ratio, π_r, versus corrected mass flow, m_c, on a pair of axes, and a K-g/ρ^* map on another pair of axes, as shown in Figure 3.3.

In the dual plots of Figure 3.3, a × is drawn on each map to fully specify the operating point, and the boundary is shown by the hatched curves. Movement of the operating point, ×, on one map affects the position of the flutter boundary on the other map. For example, if there is an increase in inlet temperature at constant corrected performance, the × moves on the K^*-g/ρ^* map, and the boundary moves on the performance map.

Conversely, for changes in corrected performance, the boundary on the K^*-g/ρ^* map moves. Figure 3.4 shows this behavior for a 2-D, subsonic, linearized-unsteady, potential flow model as described by Hall (1993) for a cascade in the 10th Standard Configuration. This model was used to compute l' in the stability criterion (3.6), for a fixed inlet flow angle, α, and varying inlet relative Mach numbers from $M = 0.4$ to $M = 0.7$.

The flutter boundary on the K^*-g/ρ^* map, for a given performance point as in Figures 3.3 and 3.4, is downward sloping. Thus, increases in g/ρ^* and in K^* are stabilizing. That increasing g/ρ^* is stabilizing can be seen by inspection of the stability criterion, (3.6). Increasing K^* is stabilizing as well, although this trend ultimately has an empirical basis. By similarity, an increase in K^* (for constant g/ρ^*, m_c, and N_c) is equivalent to an increase in the modal frequency, ω_0, whose stabilizing effect is established in design practice (Snyder and Burns, Chapter 22 of AGARD Manual, 1987). Moreover, models in many special cases exhibit this tendency, such as those for cascades in potential flow (Whitehead, 1960), inviscid cascades with strong shocks (Goldstein, et. al. 1977), and fully stalled cascades in incompressible flow (Chi, 1980). These trends in K^* and g/ρ^* can be used to assist in data interpretation and testing.

3.3 APPLICATION TO FULL-SCALE ENGINE TESTS

The above framework was applied to the interpretation of full scale engine tests. First, development tests upon an early multi-mission aircraft engine are discussed, and then a particular case of fan flutter data, provided by the Volvo corporation, from full scale engine tests is treated in detail.

Figure 3.5 shows a summary of results from flutter testing on several fan builds for a historic, multimission aircraft engine. In each case, the flutter occurred at approximately the same point on the operating map (at part speed); therefore, only the data on the K^*-g/ρ^* map is shown. Although the actual flight envelope was not available, the "generic" supersonic envelope from Figure 3.2 is used. Initial rig testing at SLS(sea-level static) (*) indicated that the engine would be stable, but engine testing at other flight conditions (cluster of O's) exhibited a flutter. Several minor modifications to the design were tested (also in the cluster of O's), but none of these increased K^* enough for stability. The final redesign, ×, did significantly increase K^*, and it is clear that the corresponding flight envelope is completely to the right of the SLS (*) point. This view shows that the stable measurements of the SLS rig test were actually consistent with the engine tests at other flight conditions, a point which was not apparent otherwise.

3.3.1 Volvo Data

The Volvo engine test data of fan flutter in the 2nd bending mode are rare in the sense that the full 4-dimensional parameter space was explored. For each combination of parameters, a frequency and magnitude of a stress signal associated with blade vibrations was recorded. The mechanical damping, g, was not explicitly measured, but it was known that there was friction at the root attachment and the part-span shroud. In the current analysis, it is assumed that the value of g was constant, and the subsequent values of g/ρ^* listed for this data are normalized relative to the sea-level static condition at $g/\rho^* = 1$.

Due to centrifugal stiffening, the rotating natural frequency, ω_{rot}, differed from the non-rotating natural frequency, ω_{rest}. In the computation of K^*, the non-rotating frequency, ω_{rest}, was used as described in the previous section, which ensures that flight condition effects and performance effects remain decoupled. However, one would expect the oscillation frequency in flutter to be ω_{rot}.

To distinguish between stable points and flutter points, both the frequency and magnitude information was used, since some of the vibrations were a result of upstream forcing. The following criteria were used to separate the data into stable and flutter data:

Stress > 1 MPa	Freq. within 2% of ω_{rot}	Category
No	-	Stable
Yes	No	Discard
Yes	Yes	Flutter

The data were arranged in clusters on the K^*-g/ρ^* map, with each cluster spanning a portion of the performance map. Each cluster did not necessarily contain enough data, alone, to obtain an accurate estimate of the stability boundary. However, data from other clusters could be used to generate upper and lower bounds for stability and thus lead to an accurate boundary. Here, we use the trend identified in the previous section that increases in g/ρ^* and in K^* are stabilizing.

Figure 3.6 shows the stability boundary estimate for the cluster, or "analysis box," with $0.68 < K^* < 0.69$, and $0.74 < g/\rho^* < 0.75$. The stability boundary is a polynomial fit using a classification algorithm (see Khalak, 2000, for details) which puts stable points on one side (lower bound) of the boundary and unstable points on the other (upper bound). The lower bound of stable points is formed from the stable points in the analysis box (*) and the stable points (+) from regions of equal or lower stability, $K^* < 0.69$ and $g/\rho^* < 0.75$. Conversely, the upper bound of unstable points is taken from measurements in the box (O) and unstable points (∇) from regions of equal or greater stability, $K^* > 0.68$ and $g/\rho^* > 0.74$.

3.3.2 Results of Volvo data analysis

The trends in the flutter boundary with changes in K^* and g/ρ^* are shown in Figure 3.7a. As expected, an increase in g/ρ^* stabilizes the flutter boundary, moving the flutter region away from the operating region. The range of movement with g/ρ^* is striking since it spans nearly half the speed-line for a change in g/ρ^* from 0.6 to 1.0. For example, at 83% corrected speed on Figure 3.7a, a change in the reduced damping, g/ρ^*, from 0.6 to 1 leads to an increase in the pressure ratio at the boundary of about 15%.

The case of changing K^* is shown in Figure 3.7b for $g/\rho^* \approx 1$, for values of K^* from 0.68 to 0.75. Increasing the compressible reduced frequency, K^*, tends to stabilize the boundary, as expected. The stabilizing effect of increasing K^* is prominent at lower corrected speeds.

In the context of corrected performance, flight condition effects are reducible to changes in the inlet stagnation pressure. For flutter stability, however, the inlet pressure alone can not always account for the flight condition effects. In the case of the supersonic aircraft of Figure 3.2, for example, there are independent variations in K^* and g/ρ^* with flight condition. To demonstrate that flight condition effects are not merely reducible to inlet pressure effects, analysis boxes at the same inlet pressure and varying inlet temperature are compared in Figure 3.8.

Instead of looking at the flutter boundaries on the performance map, we can also look at the dual view of the boundary on the K^*-g/ρ^* map. This view is useful since, for given structural quantities, it is understandable in terms of the flight envelope of the aircraft powered by the engine, as depicted in Figure 3.2. We consider boundaries on the K^*-g/ρ^* map, as shown in Figure 3.9, at a constant point on the corrected performance map (m_c, N_c), similar to those depicted in Figures 3.3 and 3.4. For each of the 11 points (•) on the K^*-g/ρ^* map, a flutter boundary was estimated on the performance map. Then, for any given performance map point, each of the 11 points could be classified as either stable or unstable depending upon its corresponding boundary location. The dashed boundaries of Figure 3.9 divide the stable •'s from the unstable •'s for a specific performance map point. In particular, the dashed boundaries correspond to a corrected speed of $N_c = 74\%$, for various critical pressure ratios, π_{cr}.

This series of boundaries gives an indication of the coupled effects of K^* and g/ρ^*. The boundaries tend to curve downward, being flatter for lower K^* and lower g/ρ^*, but steeper for higher K^* and higher g/ρ^*. The curvature in the boundaries indicates that the sensitivity changes as one moves in the parameter space. Similar behavior was observed for other values of N_c.

3.4 FLUTTER CLEARANCE

Flutter clearance refers to the testing procedure performed to ensure that the engine blades will not flutter throughout the intended operating regime. The testing environment for aeroengine flutter is harsh on the measurement equipment and the machine being tested, making it important to minimize the required testing effort.

The design of a flutter test, can be framed in terms of a *test matrix,* or a testing schedule, which spans the relevant factors. In this case, the factors are given by the similarity variables: (m_c, N_c, K^*, g/ρ^*). The requirement for flutter clearance is that *the machine must be stable for every combination of factors, (m_c, N_c, K^*, g/ρ^*), that it is designed to encounter.* Testing every permutation of factors is performed in a so-called "full factorial" test.

Although this would satisfy the requirement, it is neither practical nor essential for flutter clearance. Using the concept that flutter stability increases with increases in K^* and g/ρ^*, the test matrix can be simplified. At a given (K^*, g/ρ^*) location, a stable measurement also holds for higher values of (K^*, g/ρ^*), and an unstable measurement is relevant for lower values of (K^*, g/ρ^*). This principle is outlined in Figure 3.10.

Using this trend, we can modify the above stability requirement in the following manner: *for every performance point of operation, (m_c, N_c), the machine is stable at* (min

K^*, min (g/ρ^*)), *where* min *denotes the minimum value encountered for the given (* m_c, N_c*)*. The behavior at this minimum (K^*, g/ρ^*) pair must be estimated by some reliable means, the most straightforward of which is to take experimental measurements there.

To illustrate, a flutter clearance example is presented for a supersonic aircraft, using the Volvo engine data for the stability characteristics in non-dimensional form; however, since different structural parameters and a different flight envelope is used, the results are not related to any real case. This fictitious example is meant only to demonstrate the flutter clearance methodology, rather than to precisely simulate any specific application.

3.4.1 Example: High Speed Aircraft

We consider a front stage fan of a high-speed (i.e. supersonic) aircraft with the following assumptions, for simplicity: (a) the mechanical damping does not vary significantly during operation, (b) the relevant mode-shape has been identified.

The value of K^* is taken to be 0.8 at sea level static (SLS) conditions, and the mechanical damping remains unchanged from the Volvo data. Furthermore, we assume that SLS testing does not indicate flutter.

The minimal region for clearance on the K^*-g/ρ^* map depends upon the mission requirements, which for a high-speed aircraft are relatively sophisticated, and are based upon accomplishing specific scenarios. For this simplified example, the requirements are depicted in Figure 3.11a. Minimum required Mach numbers are specified for low altitude (take-off and climb), and for higher altitudes (cruise and supersonic dash). There are two critical points for the purposes of flutter, labeled 2' and 2", respectively. The corresponding region on the K^*-g/ρ^* map is shown in Figure 3.11b. Furthermore, for this simplified example, we require that the operating line be clear of flutter throughout the entire specified region of the K^*-g/ρ^* map. A more sophisticated approach might require a desired flutter margin, or limit the range of the operating line depending upon the values of K^* and g/ρ^*.

It is proposed to test the engine successively at the numbered points: 1, 2, and 3 on the K^*-g/ρ^* map, depicted on Figure 3.11b. At each point, the engine should be run along the operating line to check for possible intersections with the flutter boundary. This type of testing requires the capability to adjust the values of K^* and g/ρ^*, which may be accomplished by setting the temperature and the pressure of the inlet flow to achieve the desired K^* and g/ρ^*. Using the principle shown in Figure 3.10, if point 2 is confirmed to be clear of flutter, then the entire region in Figure 3.11b is clear. However, if a flutter event occurs at point 2, then the minimal region may still be cleared if points 2' and 2" are clear.

The relationship between the flutter boundary and the operating line, using the Volvo data, is depicted in Figure 3.12. By assumption, point 1 (the SLS point) does not exhibit flutter. However, point 2 would show a flutter event on the operating line. Further testing at points 2' and 2", however, would reveal that the operating line is clear throughout the minimally acceptable range of flight conditions, but that extensions beyond this minimally acceptable range (e.g. to point 2) can lead to a flutter event.

3.5 SUMMARY AND CONCLUSIONS

A framework for flutter stability was developed for the case of a machine with specified flowpath geometry and a particular mode-shape. Such a framework can be described in terms of four non-dimensional parameters, by combining the mechanical damping and the fluid inertia into a single reduced damping parameter, g/ρ^*, a novel development in the context of turbomachinery flutter. Consideration of the reduced damping implies that density effects upon flutter are only significant in the case of significant mechanical damping, usually achieved with frictional supports.

Besides corrected performance, flight condition in terms of inlet temperature and density influences flutter stability. In general, the performance can be described by m_c and N_c, while the flight condition for given modal parameters corresponds to the similarity parameters K^* and g/ρ^*. Increasing g/ρ^* has a stabilizing effect on flutter, a trend which is based in the analytical form of the stability criterion. Increasing K^* is also stabilizing, which has been shown to hold in a variety of idealized cases, but whose application to aeroengines is ultimately empirical.

The framework was used in the interpretation of full-scale flutter data, showing how one could quantify the trends in a real case. In particular, the data also exhibited the trend that increasing K^* and g/ρ^* were both stabilizing influences towards flutter. Also, it was shown that these quantities affect flutter differently, and that inlet pressure is not the only relevant inlet condition, but rather that two thermodynamic states (e.g. temperature and density) must be specified.

Finally, these concepts were applied to the problem of flutter clearance testing. The basic requirement that the machine must be confirmed to be stable at every point of operation, can be simplified by utilizing the trend that increasing K^* and g/ρ^* is stabilizing. This helps identify critical points in the required operating regime for clearance testing.

3.6 References

R.M. Chi, 1980, "Unsteady aerodynamics in stalled cascade and stall flutter prediction," ASME Paper 80-C2/Aero-1.

H. Fǫrsching, 1994, "Aeroelastic stability of cascades in turbo machinery," *Progress in Aerospace Sciences*, 30:213–266.

G.A. Gerolymos, 1993, "Advances in the numerical integration of the three-dimensional Euler equations in vibrating cascades," *Journal of Turbomcahinery*, 115:781–790.

M.E. Goldstein, W. Braun, and J.J. Adamczyk, 1977, "Unsteady Flow in a supersonic cascade with strong in-passage shocks," *Journal of Fluid Mechanics*, 83:569–604.

K.C. Hall, 1993, "Deforming grid variational principle for unsteady small disturbance Flows in cascades," *AIAA Journal*, 31:777–787.

K. Isomura, 1996, "A numerical investigation of ¤utter in a transonic fan," Technical report, MIT Gas Turbine Laboratory, GTL Report 223.

J.D. Jeffers and C.E. Meece, 1975, "F100 fan stall Flutter problem review and solution," *Journal of Aircraft*, 12:350–357.

R.R. Jutras, R.B. Fost, R.M. Chi, and B.F. Beacher, 1983, "Subsonic/transonic stall Flutter investigation of a rotating rig," Technical Report, NASA CR-174625.

R.R. Jutras, M.J. Stallone, and H.R. Bankhead, 1980, "Experimental investigation of Flutter in mid-stage compressor designs," AIAA Paper 80-0786.

J. L. Kerrebrock, 1992, *Aircraft Engines and Gas Turbines*. MIT Press.

A. Khalak, 2000, *"Parametric Dependencies of Aeroengine Flutter for Flutter Clearance Applications,"* GTL Report No. 231, Gas Turbine Laboratory, MIT.

F. Lane, 1956, "System mode shapes in the ¤utter of compressor blade rows," *Journal of the Aerospace Sciences*, 54–66.

B. McCormick, 1995, *Aerodynamics, Aeronautics, and Flight Mechanics*, Wiley.

C.M. Mehalic, H.G. Hurrel, J.H. Dicus, J.F. Lubomski, A.P. Kurkov, and D.G. Evans, 1977, " Experimental results and data format of preliminary fan Flutter investigation using YF100 engine," Technical Report NASA TM SX-3444, NASA.

W.J. Rakowski, D.H. Ellis, and H.R. Bankhead, 1978, "A research program for the experimental analysis of blade instability," In *AIAA/SAE 14th Joint Propulsion Conference*.

C. Scruton, 1965, "On the wind-excited oscillations of towers, stacks, and masts," In *Proceedings of the Symposium on Wind Effects on Buildings and Structures*, pages 798–836, London, Her Majesty's Stationary Office.

F. Sisto and F.O. Carta (ed), 1988, *AGARD Manual on Aeroelasticity in Axial-Flow Turbomachines Volumes 1 and 2* AGARDograph No. 298.

A.V. Srinivasan, D.G. Cutts, and S. Sridhar, 1981, "Turbojet engine blade damping. Technical report, NASA Contractor Report, No. 165406.

A.V. Srinivasan, 1997, "Flutter and resonant vibration characteristics of engine blades," *ASME Journal of Engineering for Gas Turbines and Power*, 119:742–775.

H. Stargardter, 1979, "Subsonic/transonic stall Flutter study. Technical report NASA CR-165356.

J.M. Verdon., 1993, "Review of unsteady aerodynamic methods for turbomachinery aeroelastic and aeroacoustic applications", *AIAA Journal*, 31:235–250.

B.J. Vickery and R.D. Watkins, 1964, "Flow-induced vibrations of cylindrical structures," In R. Silvester, editor, *Proceedings of the First Austrailian Conference on Hydraulics and Fluid Mechanics*. Pergamon Press.

D.S. Whitehead, 1960, "Force and moment coeff£cients for aerofoils vibrating in cascade. Technical Report R.&M. 3254, British A.R.C.

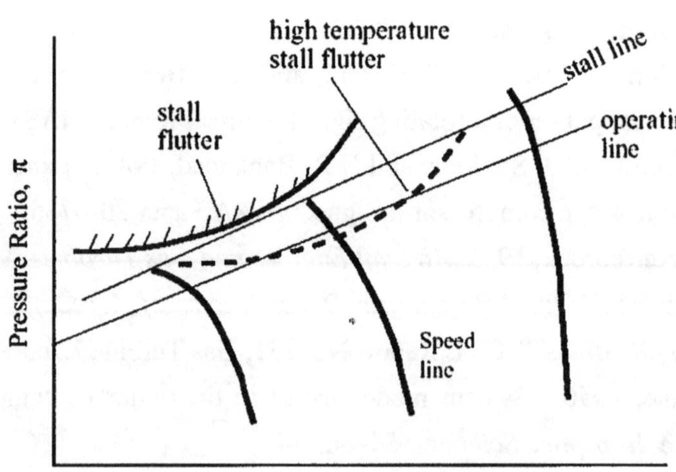

Fig. 3.1 Schematic of performance map with stall flutter boundary. Changes in the thermodynamic conditions can move the boundary, as shown.

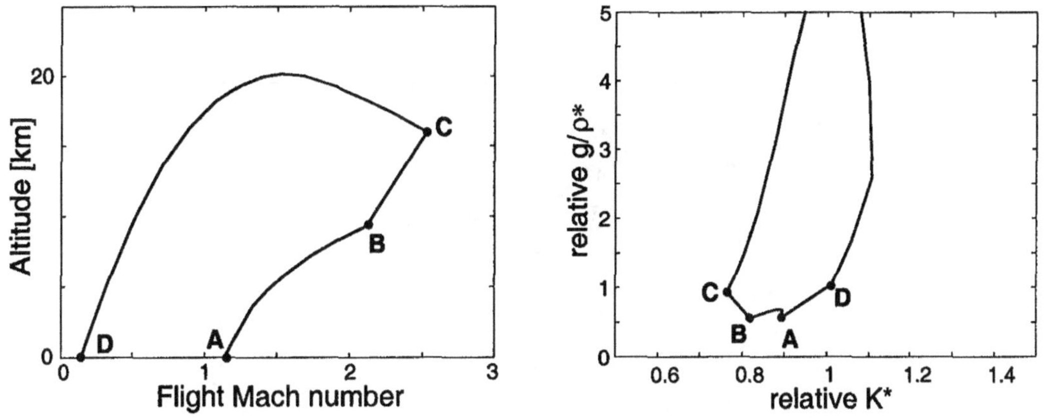

Fig. 3.2 Flight envelopes for typical supersonic aircraft of flight Mach number versus altitude. The corresponding plot of (K*, g/ρ*) is shown, assuming a front stage with an ideal inlet and constant modal parameters. The (K*, g/ρ*) plot is normalized such that sea-level static conditions are at point D, location (1,1).

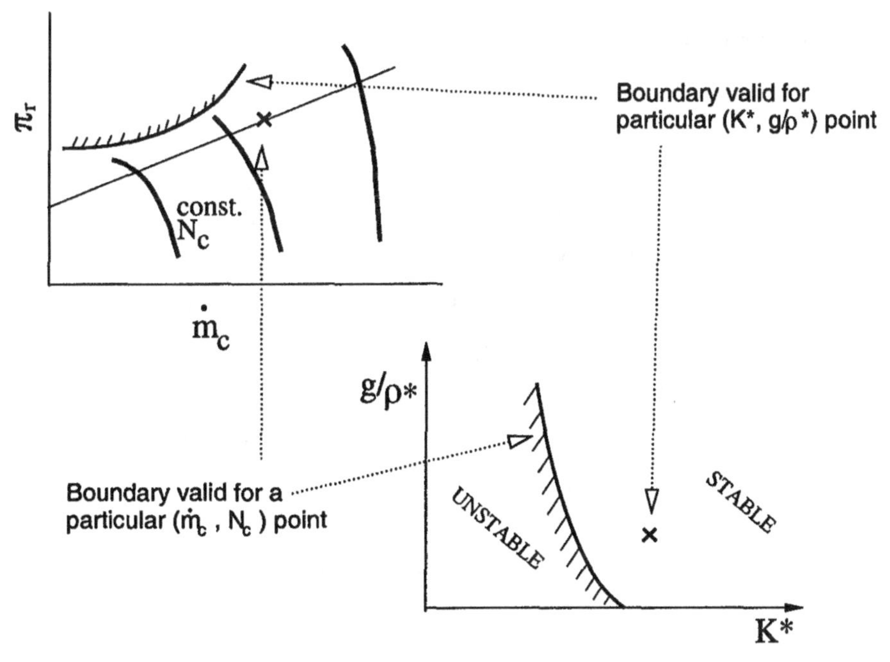

Fig. 3.3 Dual view of performance map and K*-g/ρ* map. Two simultaneous views depict a point (denoted by the x) in relation to the flutter boundary in the four parameter space, (m_c, N_c, K*, g/ρ*). Movement of the x on one set of axes affects the flutter boundary location on the other set of axes. A full description of the operating point requires an x on both axes.

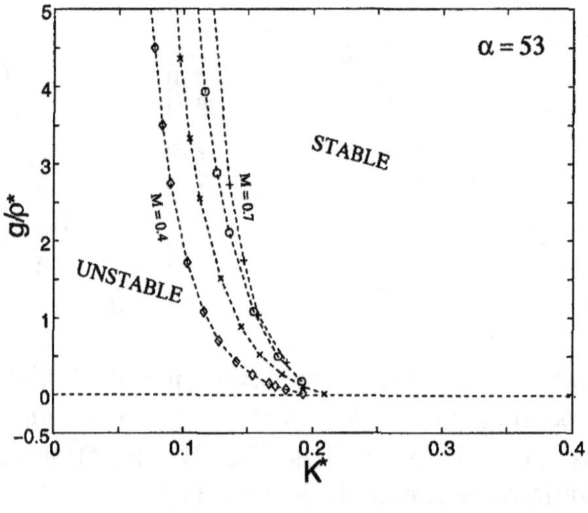

Fig. 3.4 Boundaries on K*-g/ρ* map using linearized unsteady compressible potential model in 10th Standard Configuration. The effect of increasing Mach number at constant (K*, g/ρ*) is destabilizing.

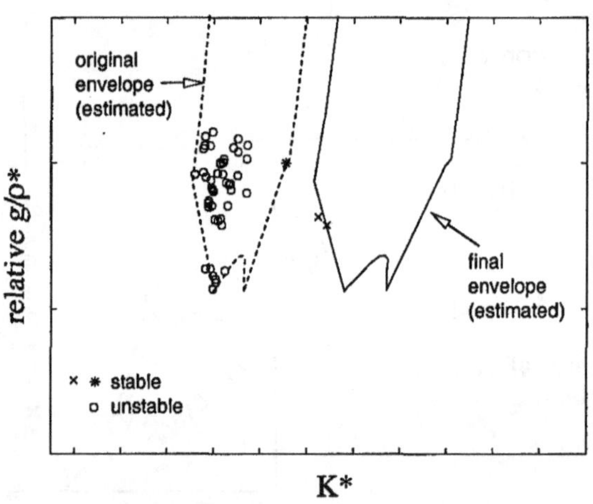

Fig. 3.5 Fan data on K*-g/ρ* map, for early multimission aircraft. A o indicates a flutter point, while * and x are stable points. The dashed envelope is estimated from the "generic" aircraft shown in Fig. 2, anchored on the sea level static (SLS) rig tests on the original design, *. The original design (and several minor design variations) were unstable at flight conditions other than SLS, shown in the cluster of o's. The eventual redesign, corresponding to the solid envelope, was tested to be stable at all relevant flight conditions, x.

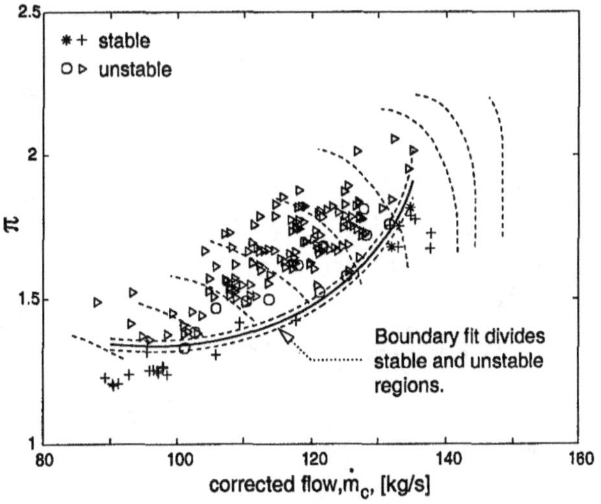

Fig. 3.6 Example of stability boundary, for (0.68 <K*<0.69 and 0.74 <g/ρ*<0.75), with o in flutter, and * stable. Data outside this (K*, g/ρ*) analysis box, + points at lower K* and g/ρ*, and ∇ at higher K* and g/ρ*, are used to generate upper and lower bounds for the stability curve. The dashed lines indicate the uncertainty in the boundary estimation process.

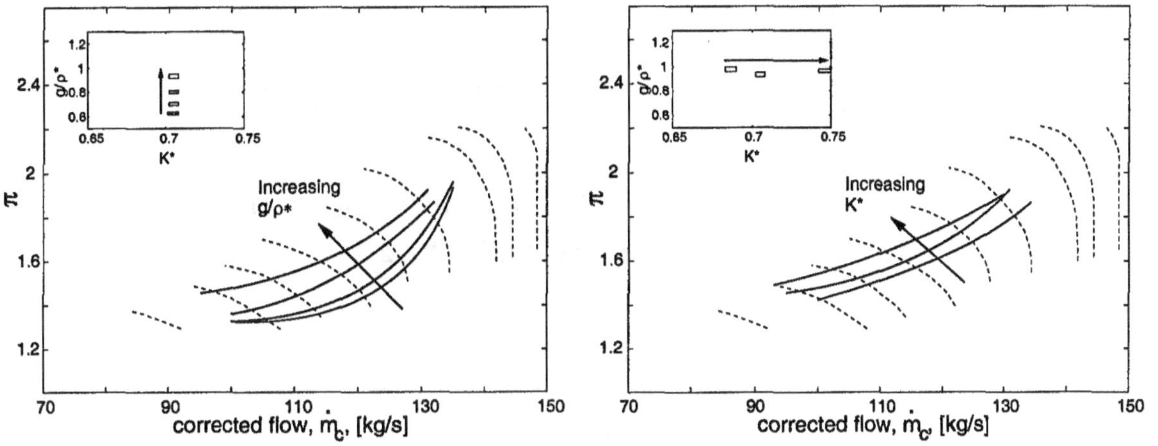

(a) (b)

Fig. 3.7 Flutter boundaries (a) with constant K*=0.705, and varying g/ρ*, and (b) with constant g/ρ*=1, and varying K*. Each boundary corresponds to one of the analysis boxes of the inset plots. The trend for increasing g/ρ*, is stabilizing, as is the trend for increasing K*

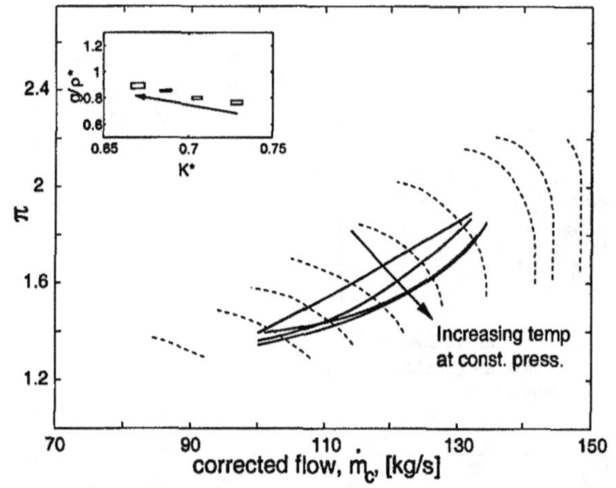

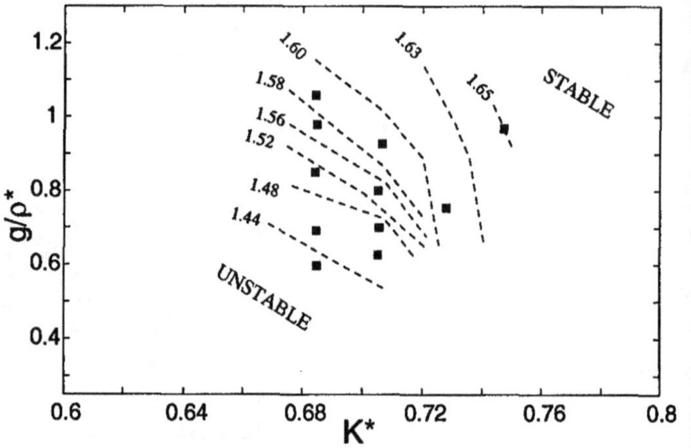

Fig. 3.8 Effect of temperature at constant pressure. A series of flutter boundaries are shown corresponding the same pressure (within 2%) and varying temperature. Increasing in temperature, at constant pressure, destabilizes flutter boundary, showing that inlet pressure is not the only relevant flight condition.

Fig. 3.9 Family of flutter boundaries K*-g/ ρ* map, for 74% corrected speed, and various critical pressure rations, π_{cr}. The boundary resolution is limited by sampling of (K*, g/ρ*) points (represented by ■'s) .

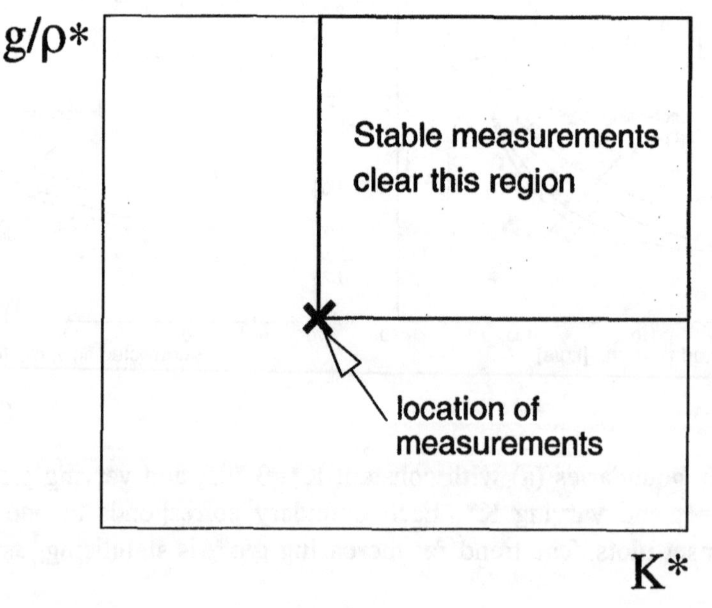

Fig. 3.10 Schematic of flutter clearance rule on K*-g/ρ* map

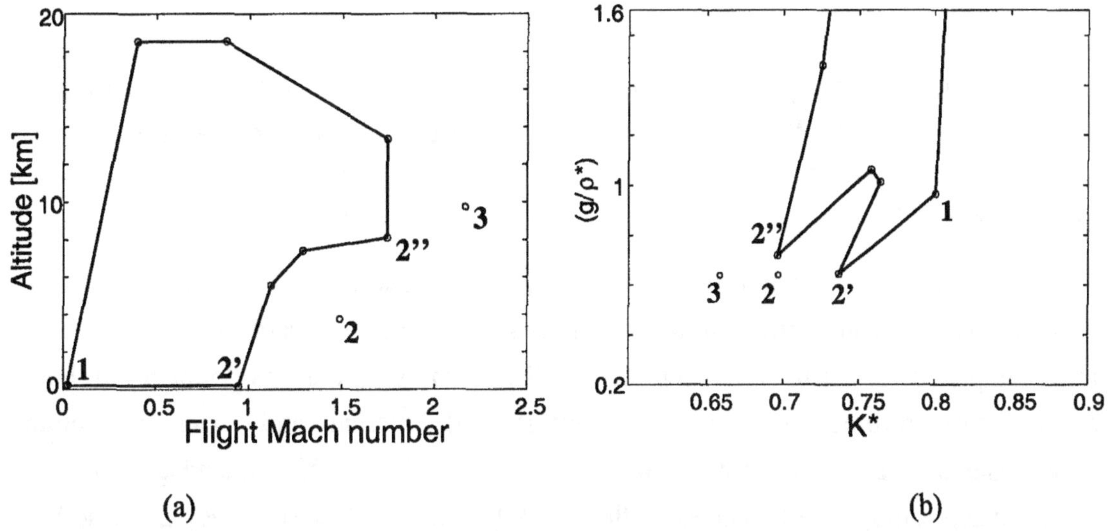

(a) (b)

Fig. 3.11 Flight requirements in terms of Mach number versus altitude (a) and region on K*-g/ρ* map (b). The sea level static condition is assumed to be at (K*, g/ρ*)=(0.8,1). The critical points, 2' and 2'' are labeled on both plots

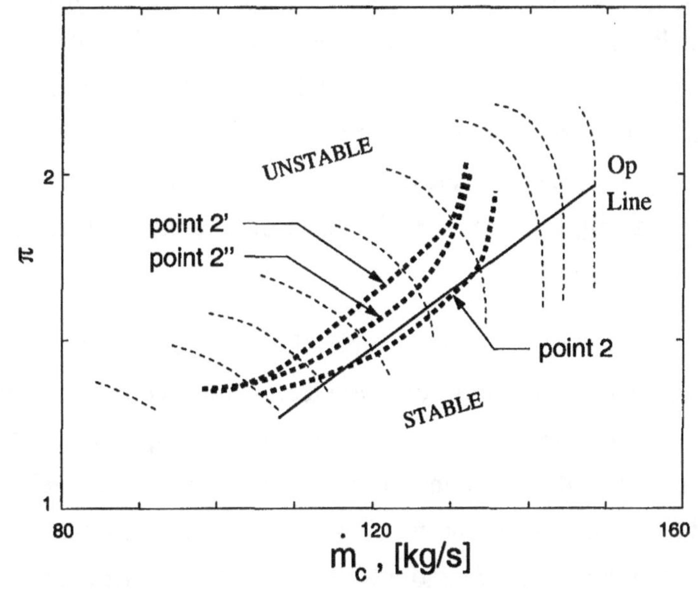

Fig. 3.12 Flutter boundaries for clearance example. The tests are performed at 4 points: first, at sea level static (point 1), then at point 2, where a flutter occurs on operating line, then at points 2' and 2'', which establish that the minimally acceptable envelope is clear, but is close to a flutter event on operating line at point 2.

4 PHYSICS OF COOLANTS/MAIN FLOW INTERACTIONS IN AIRCRAFT AND ROCKET TURBINES **(A. Epstein, G. Guenette)**

4.1 Introduction

To meet the demand for higher power densities from gas turbine systems, designers have devoted much effort over the past half-century toward increasing turbine inlet temperatures. This has been accomplished through the development of both extraordinary turbine materials and sophisticated internal and external cooling arrangements. The aerodynamic performance of turbines has likewise increased enormously, leading to efficiencies in the low 90% range. This efficiency increase has come through improved design techniques based on a better understanding of the fluid mechanics of turbines and the application of increasingly more powerful computational tools.

Supporting these improvements are empirical observations acquired through many years of testing on engines, component test rigs, and sub-scale experiments. Each of these test options has their advantages and limitations. For an engine test, the measurement uncertainty for individual component efficiencies is high, while sub-scale experiments may not simulate all relevant physical phenomena. An option is the full-scale steady state test rig, however the high cost of running such facilities has limited turbine research at this level.

Over the past two decades a technology based on short-duration testing techniques has been developed. Short duration tests can provide highly accurate, detailed turbine measurements at relatively low cost. The methodology was made possible by the rapid advancement of high frequency response instrumentation and data acquisition technologies and by a realization that the physically relevant time scales of the processes within a turbine are on the order of hundreds of microseconds. Under suitable conditions, test durations of less than a second can provide tens of thousands of characteristic flow times over which to conduct an experimental investigation. Test facilities based on these methods can provide significant benefits to measurement resolution and flexibility while reducing the costs of the test hardware and operation.

Short duration techniques have subsequently been developed and successfully applied by a number of investigators to the detailed study of cooled turbine heat transfer. Methods have also been developed Keogh (2000) to measure turbine aerodynamic performance in short duration environments. The aim of the present work is to utilize these techniques to provide a detailed study on the influence of film cooling on the aerodynamic performance of a fully scaled film cooled turbine stage. The work presented herein covers the experimental results obtained with a film-cooled turbine stage and a corresponding uncooled stage.

4.2 Experimental Approach

The tests were conducted in the MIT Blowdown Turbine Facility which is a short duration wind tunnel capable of testing a complete rotating stage in an environment which fully simulates the fluid physics and heat transfer phenomena of an operational turbine. The overall configuration and operational procedures of the facility are briefly reviewed as well as the instrumentation and coolant feed system. The scaling of the test conditions to achieve full-scale similarity is discussed.

4.2.1 Facility Description

The facility, shown in Figure 4.1, is composed of seven primary components: the supply tank, fast acting valve, test section, eddy current brake torque meter, critical flow venturi nozzle, coolant feed system, and dump tank. A detailed review of the overall design can be found in Guenette (1985).

Supply Tank and Valve The supply tank is a 364 cubic foot pressure vessel surrounded by a circulating oil heating system to achieve the desired inlet gas temperature. It is separated from the test section by a fast acting (~50ms) plug valve that introduces a disturbance free flow into the test section entrance annulus.

Test Section The test section consists of the forward frame, main frame and rotor unit as shown in Figures 4.2 and 4.3. The forward frame contains the nozzle guide vanes, the nozzle coolant plenum and the inlet boundary layer bleed, and mates with the fast acting valve. The forward frame is followed by the main frame which houses the turbine rotor assembly, the downstream probe translator, the eddy current brake magnet and torque measuring load cells, the turbine throttle and drive motor. The rotor unit contains the disks, blades, bearings, shaft, and the eddy current brake loss drum, (not shown in Figure 4.3). As shown in Figure 2, the drum is inserted into the eddy current magnet assembly.

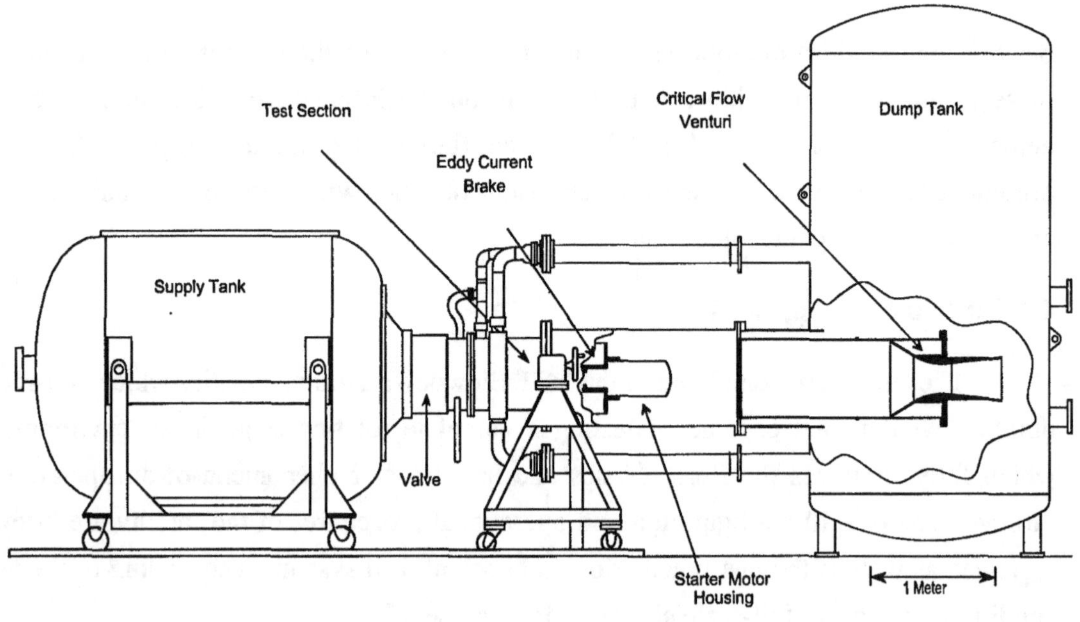

Fig. 4.1 MIT Blowdown Turbine Facility

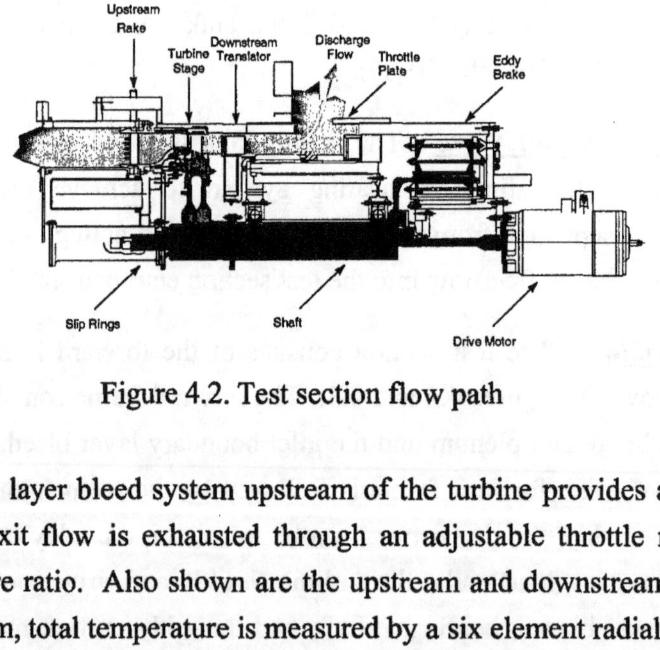

Figure 4.2. Test section flow path

A boundary layer bleed system upstream of the turbine provides a uniform inlet flow. The rotor exit flow is exhausted through an adjustable throttle ring to set the desired test pressure ratio. Also shown are the upstream and downstream measurement locations. Upstream, total temperature is measured by a six element radial rake and three center-annulus probes spaced equally around the circumference. A five element radial rake and three circumferential probes provide total pressure measurements.

The downstream translator houses three canisters, spaced twenty degrees apart, which contain the appropriate connections to support either pressure or temperature measurement instrumentation. Connections are also provided for several wall static pressure taps. For these tests an eight element total pressure rake, a four element total temperature rake and a five element total temperature rake were employed. The translator can be set to perform a programmed circumferential sweep during the test interval.

Eddy Current Brake Torque Meter The eddy current brake (figure 4.4) serves as both the turbine load and turbine torque meter. It consists of two components: 1) a cylindrical, INCONEL drum directly attached to the shaft and 2) a stationary array of DC excited electromagnets arranged circumferentially around the drum. Electric currents, induced by the motion of the drum through the applied magnetic field, lead to resistive heating of the drum and dissipate the power produced by the turbine. The brake load is controlled by the magnitude of the applied magnetic field. The braking system was modified by Keogh (2000) to serve as a shaft torque meter. The electromagnet assembly was mounted on bearings and two load cells measured the force required to restrain the assembly. Figure 4 shows a cross section through the brake.

Critical Flow Venturi Mass flow measurement was accomplished by the installation of an ASME/ANSI (1987) standard toroidal throat critical flow venturi nozzle downstream of the turbine throttle. The nozzle and its required entrance duct were installed in the facility dump tank. An extension was added to the dump tank inlet to accommodate an inlet conditioning duct and a 50% open area screen to smooth the flow transition from the annular turbine exit to the circular venturi inlet. The system was

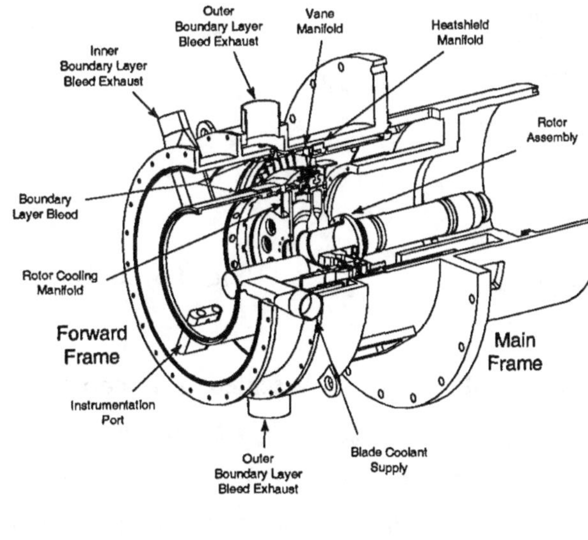

Fig. 4.3 Test section detail

designed and built by Flow Systems Inc. of Bolder Colorado and calibrated by Colorado Experimental Station Incorporated. The calibration is traceable to the National Institute of Standards and Technology. More detail on its design and usage can be found in Keogh (2000).

Coolant Feed System The coolant feed system, Spadaccini (1999), consists of a liquid nitrogen chilled supply tank, a fast acting pneumatic ball valve, and several mass flow metering orifice plates. Like the main facility, the cooling system operates in a blowdown mode. It provides three independent and separately metered flows to the vanes, blades, and rotor tip casing. Figure 4.5 shows a schematic of the cooling system, indicating the locations of the pressure and temperature instrumentation and each metering orifice.

4.2.2 Film-Cooled Turbine Stage

A ¼-scale film cooled turbine stage was manufactured maintaining the external profile, film cooling hole quantity, location, and injection angle. The internal passages were not directly replicated as the focus of the program was on external aerodynamics. Electro-discharge and laser machining techniques were used to fabricate the film-cooled airfoils, Spadaccini (1999).

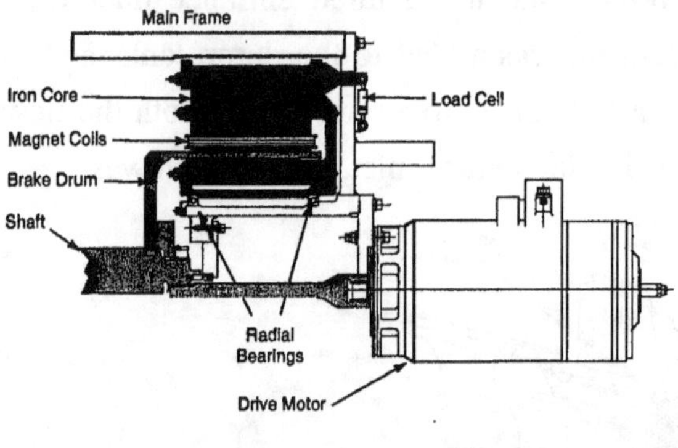

Fig. 4.4 Eddy current brake.

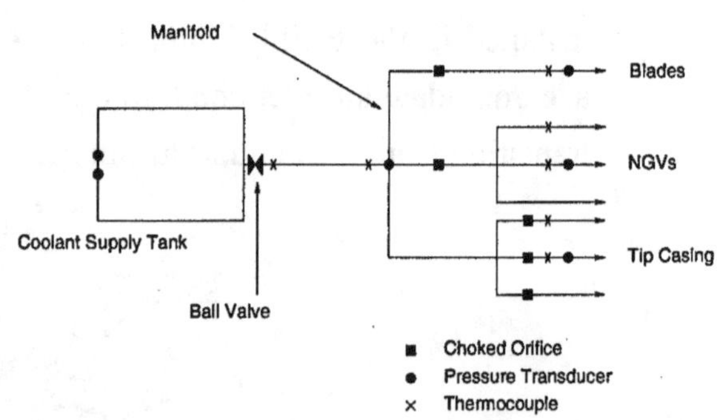

Fig. 4.5 Coolant feed system schematic.

Film-Cooled Nozzle Guide Vanes A scaled, film-cooled nozzle guide vane is shown in Figure 4.6. The vane cooling configuration consists of 12 rows of coolant injection holes on the airfoil surface, as well as holes on the upper and lower platforms and trailing edge. Two internal channels were required to feed the cooling holes. Coolant was supplied via manifolds surrounding the hub and tip platforms.

Film-Cooled Rotor Blades The scaled film-cooled rotor blade is also shown in Figure 4.6. The blade cooling configuration consists of four rows of coolant injection holes: two on the leading edge, one on the suction surface, and one on the pressure surface. The leading edge holes are also oriented at compound angles. Two channels were used to feed the blade cooling holes, one for the leading edge and suction surface rows and one for the pressure surface row. The blade plenums were sealed at the blade tips by laser welded caps.

Film-Cooled Rotor Tip Casing The engine rotor tip casing configuration could not be directly incorporated into the MIT test facility. As a result, a geometry was chosen that provided the same coolant mass flow rate but with a modified injection distribution.

4.2.3 Scaling and Test Conditions

In dimensionless form, the equations for mass, momentum and energy plus the associated geometric boundary conditions produce non-dimensional parameters that completely govern the flow. Similarity between experimental and full-scale flow only requires that these non-dimensional parameters be reproduced. For the adiabatic un-cooled turbine, corrected mass flow and total temperature ratio depend on total pressure ratio, corrected speed, gas specific heat ratio, and Reynolds number. Two additional parameters are required to simulate the heat flux distribution, the gas to wall temperature ratio and the Prandtl number. For the film-cooled turbine, the coolant mass flux and momentum flux ratios must be reproduced.

These governing parameters can be matched by properly setting the facility initial operating conditions. The operating conditions include the test gas composition, the turbine inlet total pressure and temperature, the coolant inlet total pressure and temperature, the mechanical speed, the throttle position, the coolant metering orifice diameters, and the brake excitation. The supply tank temperature is set by matching the gas to wall temperature ratio. Specific heat ratio is set by the gas composition, typically a mixture of argon and carbon dioxide. The shaft speed is set to match the turbine corrected speed. Initial supply pressure and throttle position are set to match the desired Reynolds Number and the stage pressure ratio.

4.3 Results

Representative data for the aerodynamic performance experiments are presented. Although dimensional quantities vary significantly over the test window, the non-dimensional parameters remain constant to within several percent. The test conditions for the film-cooled and un-cooled experiments are presented and the data reduction procedures are outlined. A performance comparison, with and without film cooling is presented. For the un-cooled configuration, the performance of the turbine stage is also compared for two rotor tip gaps.

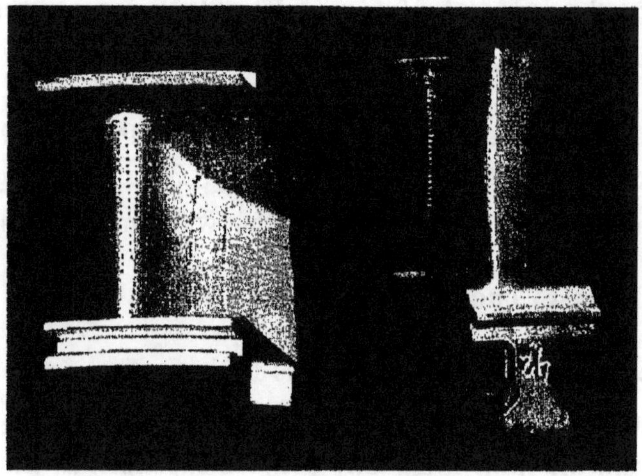

Fig. 4.6 Film-cooled NGV and blade

The turbine efficiency is defined as the ratio of the turbine power to ideal power,

$$\eta = P/P_{ideal} \tag{4.1}$$

Turbine power is obtained from the measurements of shaft speed and brake toque,

$$P = T_{brake}\ \Omega + I \frac{d\Omega}{dt} \Omega$$

Figures 4.7 and 4.8 show the mechanical speed and the brake torque measurements for a typical blow down run. Figure 4.9 shows the brake and turbine power, and illustrates the influence of rotational inertia on the results because of the shaft acceleration. The ideal turbine power is calculated from the turbine and coolant mass flows and from a computed ideal total enthalpy drop across the turbine. As the test gas exhibits significant non-ideal behavior over the experimental conditions, use of a calorically perfect gas model would under-predict the efficiency by several per-cent. Therefore real gas thermodynamic properties must be employed and are obtained from the *NIST Mixture Property Database*, NIST14 (1992).

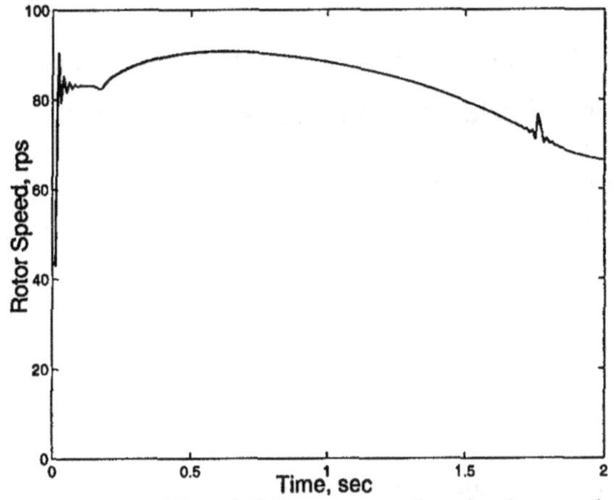

Fig. 4.7 Rotor mechanical speed.

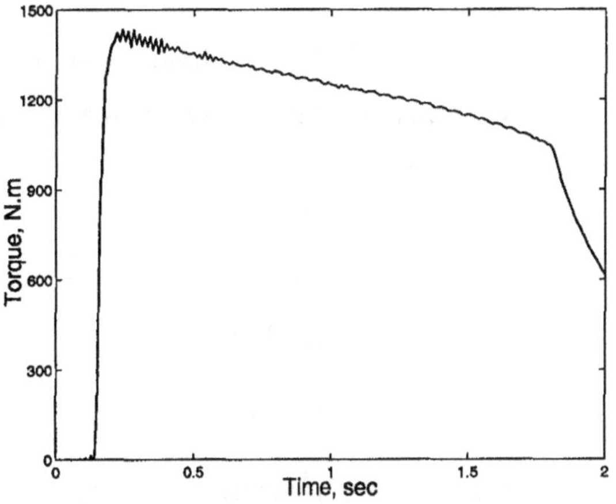

Fig. 4.8 Brake torque.

The ideal enthalpy change across the un-cooled stage is calculated using,

$$P_{ideal} = \dot{m}_t \left[h\left(T_{t,in}, P_{t,in}\right) - h\left(T^{is}_{t,out}, P_{t,out}\right) \right] \tag{4.3}$$

In this and the following equation, T and P refer to the total conditions and the subscript, *t*, refers to the turbine main flow.

The inlet enthalpy is evaluated from the NIST14 tables using the measured inlet pressure and temperature. The ideal outlet enthalpy is determined from the measured outlet pressure and the 'isentropic' outlet temperature, which is defined as that temperature at which the outlet entropy is equal to the inlet entropy. It is obtained by an iteration on the NIST14 tables.

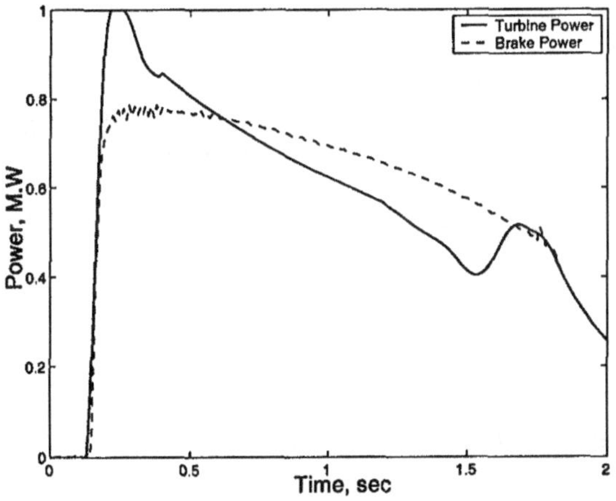

Fig. 4.9 Turbine and brake power.

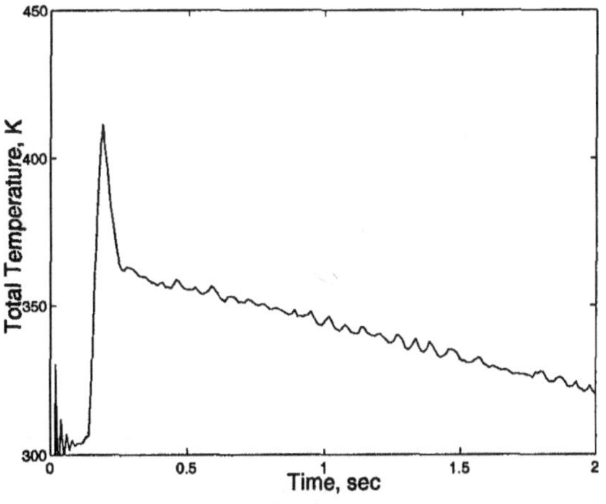

Fig. 4.10 Turbine inlet temperature.

The ideal enthalpy change across the film-cooled turbine stage is the mass weighted ideal enthalpy change of the mainstream, nozzle, rotor and tip casing streams:

$$
\begin{aligned}
P_{ideal} = & \left(\dot{m}_{t,out} - \dot{m}_n - \dot{m}_r - \dot{m}_h\right)\left(h\left(T_{t,in}, P_{t,in}\right) - h\left(T^{is}_{t,out}, P_{t,out}\right)\right) \\
& + \dot{m}_n \left(h\left(T_{n,in}, P_{n,in}\right) - h\left(T^{is}_{n,out}, P_{t,out}\right)\right) \\
& + \dot{m}_r \left(h\left(T_{r,in}, P_{r,in}\right) - h\left(T^{is}_{r,out}, P_{t,out}\right)\right) \\
& + \dot{m}_h \left(h\left(T_{h,in}, P_{h,in}\right) - h\left(T^{is}_{h,out}, P_{t,out}\right)\right)
\end{aligned} \tag{4.4}
$$

where, *n*, *r*, *h* refer to the nozzle, rotor, and tip casing coolant flows, respectively.

Turbine outlet mass flow is obtained by correcting the mass flow obtained from the critical flow venturi to account for the rate of change of mass storage in the ducting, *j*, between the turbine and the venturi throat,

$$
\dot{m}_{t,out} = \dot{m}_{cfv} + \sum V_j \frac{d\rho_j}{dt} \tag{4.5}
$$

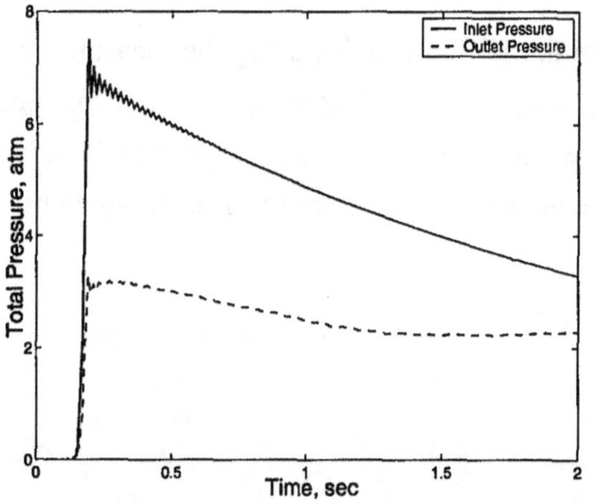

Fig. 4.11. Turbine inlet and outlet total pressure.

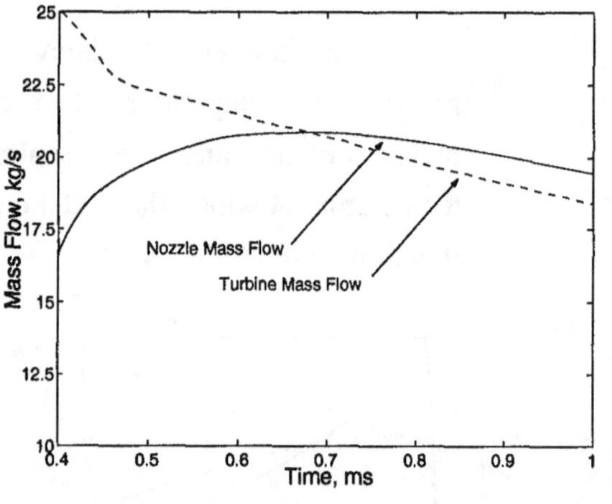

Fig. 4.12 Venturi nozzle and turbine mass flow.

Pressure and temperature measurements in ducting between the turbine and venturi nozzle are used to obtain the stored mass.

Upstream of the turbine, the mass flow for each of the coolant streams is metered using a choked orifice and a similar correction for the mass storage is employed,

$$\dot{m}_i = \dot{m}_{orifice,i} - V_i \frac{d\rho_i}{dt} \qquad (4.6)$$

where, $i = n, r, h,$ refers to nozzle, rotor, or tip casing coolant streams.

The turbine inlet temperature, inlet and outlet total pressures, and outlet mass flow are shown in Figures 4.10-4.12. Figures 4.7 through 4.12 show that the dimensional parameters vary significantly over the test window. In contrast, the non-dimensional parameters, as shown in Figures 4.13 through 4.16, remain constant for a significant period. For the test window of 0.6 to 0.9 seconds, the pressure ratio, Figure 4.13, remains within +/- 1%, the corrected speed, Figure 4.14, within +/- 2%, and the coolant mass flow ratios, Figure 4.15, within +/- 4%. As the through-flow time for the turbine is ~5 milliseconds, there is approximately two orders of magnitude more test time than required for steady operation of the turbine.

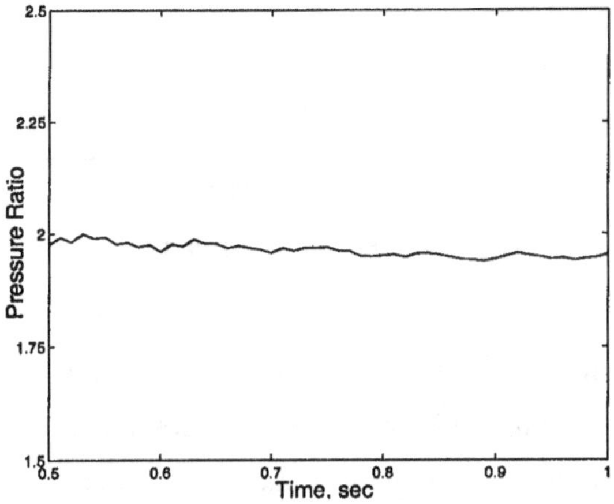

Fig. 4.13 Turbine total pressure ratio.

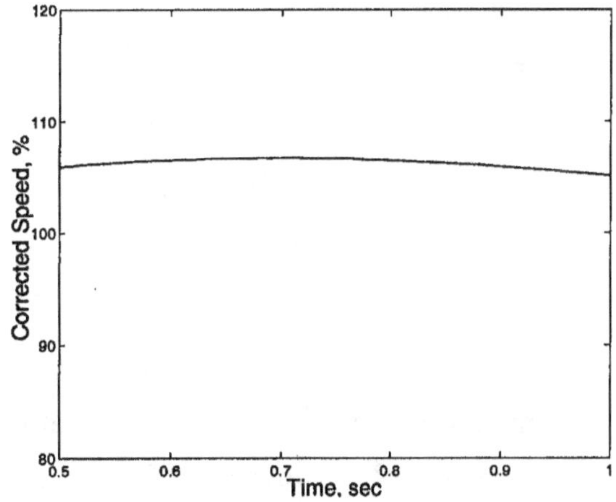

Fig. 4.14 Turbine corrected speed.

Figure 4.17 shows the pressure ratios and corrected speeds at which the un-cooled turbine was tested (with the 50% of reference tip gap). The un-cooled stage (with 100% tip gap) and the film-cool stage (with 100% tip gap) were tested at conditions as close to these as possible. As the operating points for each of the test series do not overlap, a quadratic least squares fit of efficiency was used to compare the changes with respect to pressure ratio and corrected speed.

Figure 4.18 shows the efficiency as a function of corrected speed for the un-cooled turbine using the 100% and the 50% tip gaps. In order to simplify the presentation of the data, the points shown in Figure 4.18 are the measured data as a function of corrected speed adjusted to a reference pressure ratio.

At 100% corrected speed the stage efficiency increases by 2.5%, for a rotor tip gap reduction of 50%. The measured sensitivity to tip gap is a 1.7% drop in efficiency per percent of span. The measured sensitivity varies with operating point. As shown in Figure 4.18 the sensitivity is higher at high-corrected speeds. This result is not consistent with published tip gap correlations (Horlock (1973)), most of which predict that the tip gap loss increases with higher loading (reduced speed). This result will require further investigation.

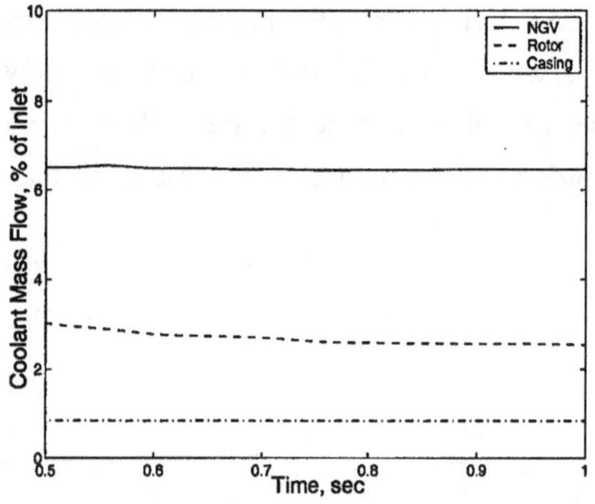

Fig. 4.15. Coolant mass flow

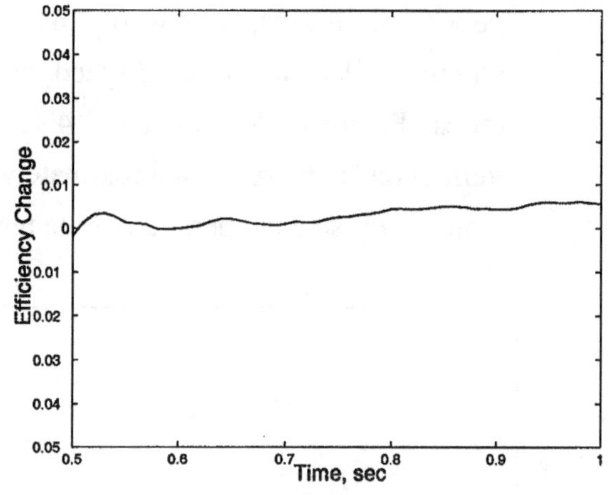

Fig. 4.16 Isentrpoic efficiency.

Figure 4.19 shows the variation of stage efficiency with coolant-to-mainstream mass flow ratio for two coolant temperature ratios. At the reference operating point and with a cooling flow of 10%, the film-cooled turbine efficiency is reduced by 1.75% compared to the un-cooled stage. This corresponds to an overall sensitivity of 0.175% per percent of cooling flow. At the reference coolant mass flow, the stage efficiency is reduced by 0.5% for each additional 1.0% of coolant added. The sensitivity increases at increased cooling flows and for lower coolant temperature (low density ratio) ratios.

Figure 4.20 shows the measured efficiencies for the two temperature ratios tested as a function of momentum flux ratio. This comparison indicates that changes in efficiency scales with momentum flux ratio.

In Figure 4.21, the results of a series of tests to examine the influence of turbine inlet temperature, or heat load, on the isentropic efficiency of the film-cooled stage can be seen. For these tests the inlet Reynolds number and the vane coolant-to-mainstream momentum flux ratio were held constant as the inlet temperature was varied. Within the uncertainty of the measurements, no variation in efficiency is observed as the heat load is varied.

The variation of stage corrected mass flow with coolant-to-mainstream mass flow ratio for the two coolant temperature ratios tested is shown in Figure 4.22. For each additional 1% of coolant the stage corrected mass flow[1] increases by 0.5%. The corrected mass flow for the low coolant-to-mainstream temperature ratio is reduced by approximately 1.0%.

Figure 4.23 shows how the stage corrected mass flow varies with corrected speed for the baseline and film-cooled turbines. The trends are significantly different; the corrected mass flow does not plateau as the corrected speed is reduced in the film-cooled tests. These trends were examined with the aid of an analytical model of the stage in Keogh (2001).

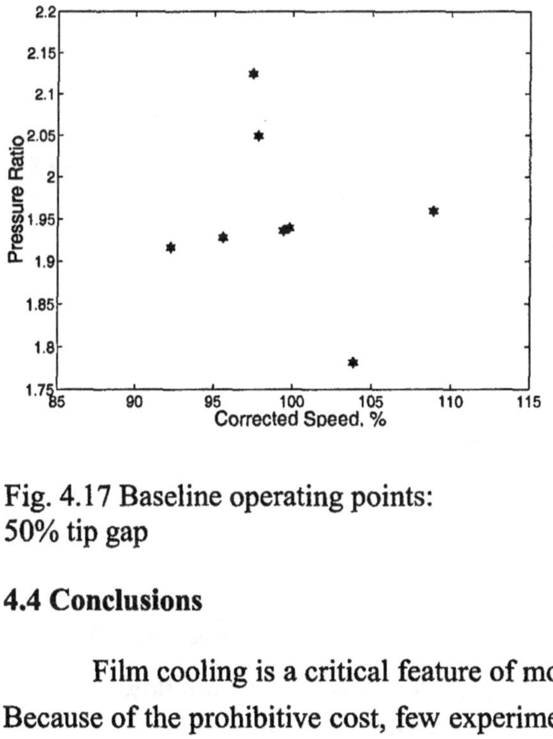

Fig. 4.17 Baseline operating points: 50% tip gap

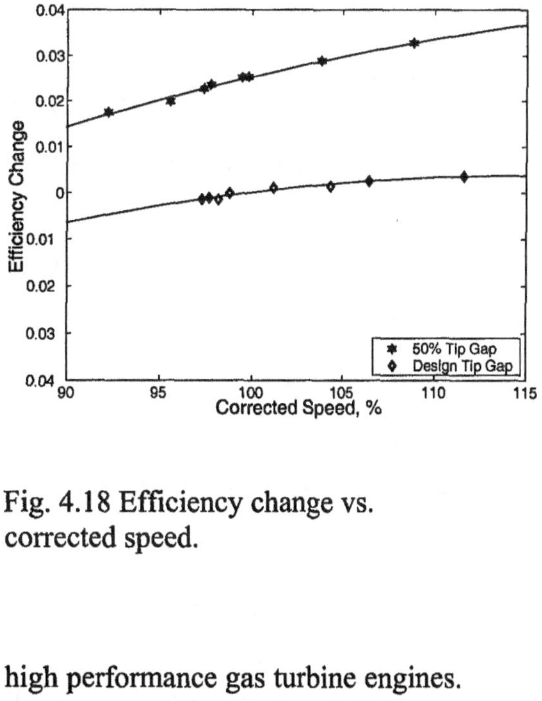

Fig. 4.18 Efficiency change vs. corrected speed.

4.4 Conclusions

Film cooling is a critical feature of modern high performance gas turbine engines. Because of the prohibitive cost, few experimental studies have measured the influence of film cooling on the aerodynamic performance of a turbine stage. Newly developed short duration measurement techniques have enabled such studies.

[1] Based on inlet total pressure and temperature and outlet mass flow.

The performance of a un-cooled stage was measured and the influence of the rotor tip gap quantified. The stage efficiency was reduced by 1.7% for each percent increase in rotor tip gap.

For the film-cooled stage, the stage efficiency was reduced by 1.75% (at the reference coolant mass flow and temperature ratio) when compared to the baseline. Each additional percent of cooling flow reduced the stage efficiency by 0.5%.

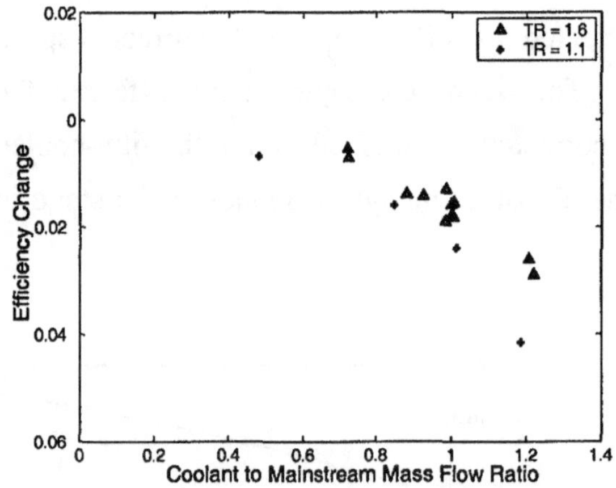

Fig.4.19 Efficiency change vs. coolant mass flow.

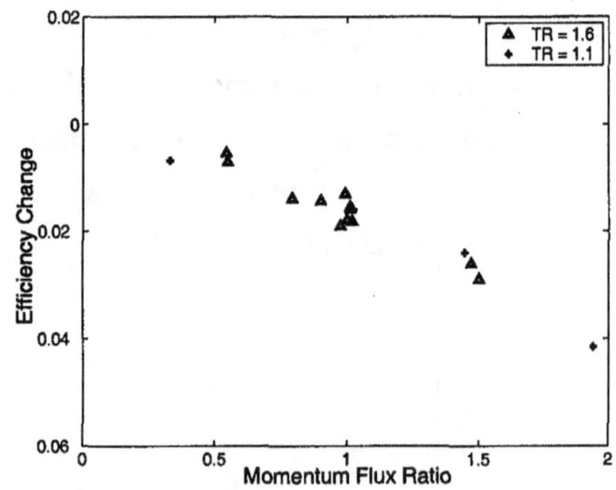

Fig. 4.20. Efficiency change vs. momentum flux ratio.

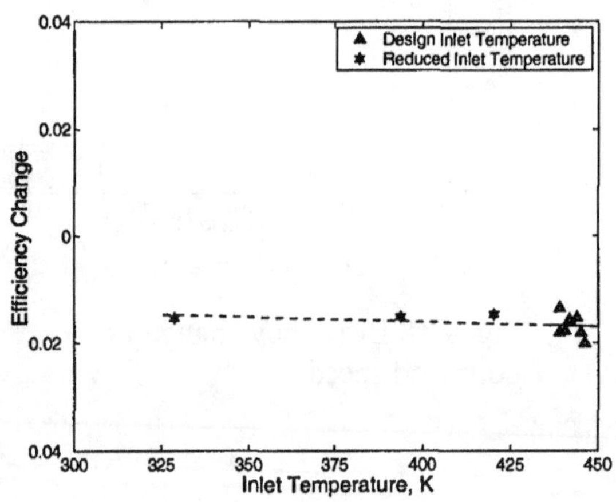

Fig. 4.21 Efficiency change vs. turbine inlet temperature

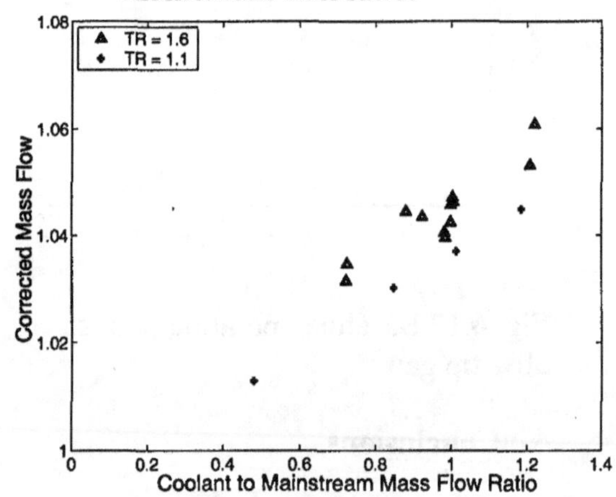

Fig. 4.22 Corrected mass flow vs. coolant mass flow.

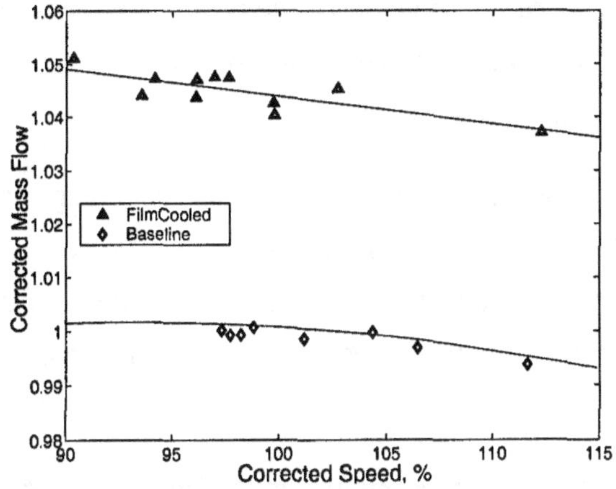

Fig. 4.23 Corrected mass flow vs. corrected speed.

Within the measurement uncertainty of the data, the increasing loss with coolant mass flow scaled with coolant-to-mainstream momentum flux ratio. Also, no change in stage efficiency was observed when the mainstream-to-metal temperature ratio was varied. These experimental observations can be used to extrapolate the results to conditions beyond those tested.

The measured change in efficiency between the un-cooled and the film-cooled stages does not represent the loss due to film cooling alone. Without film cooling the un-cooled stage is operating off design. The injected coolant flow is required tomaintain the desired stage velocity triangles and airfoil Mach number distributions.

Additional analysis was performed to account for these losses and may be found in Keogh 2001, where a meanline velocity triangle model of the stage and the film-cooed efficiency measurements are used together to estimate the loss due to film-cooling. The nozzle guide vane and rotor blade row loss characteristics were estinated as follows:

- MISES code [Drela, 1989] was used to calculate the airfoil profile losses and an additional loss was added to model unsteady Von Karman type vortex shedding behind the blunt trailing edges of the vane and blade.
- The end-wall losses were estinated using a semi-empirical loss correlation proposed by Sharma and Butler[1987].
- The influence of film-cooling and trailing edge injection on the mainstream Mach number and loss is modeled using Harstel's mixing layer approach[1972].

The meanline velocity triangle model was calibrated against the efficiency measurements of the un-cooled geometry. The measured corrected mass flow for the baseline and film-cooled tests were also used to verify the modeling assumptions.

The model was then used to estimate the performance of an ideal stage with the same velocity triangles and airfoil Mach number distributions as the film-cooled stage, but without the coolant. The estimated loss due to film cooling was found from this approach to be 3.0% at the design point, which corresponds to 0.3% per percent of coolant flow.

The loss due to film cooling is small when compared to the overall benefit that is achieved from operating the gas turbine engine at higher inlet temperatures. However, it is a significant source of loss that must be considered in detail to accurately predict the overall efficiency and power of a gas turbine engine. Keogh 2001 presents both the experimental methodology to make film-cooled efficiency measurements and analytical approach to estimate the film cooling loss from the measurements. The research has significantly reduced the uncertainty associated with predicting the film-cooling loss for the test stage.

4.5 References

ASME/ANSI MFC-7M-1987, 1987, " *Measurement of Gas Flow by Means of Critical Flow Venturi Nozzles,*" , The American Society of Mechanical Engineers.

J.D. Denton. *Loss Mechanisms in Turbomachines. ASME J. of Turbomachinary,* 115:621-656, January 1993.

M. Drela, 1989, "An Integral Boundary Layer Formulation for Blunt Trailing Edges," AIAA, 89-2200.

G.R. Guenette, 1985, " *A Fully Scaled Short Duration Turbine Experiment,*" Ph.D. Thesis, Massachusetts Institute of Technology.

J.E. Harstel, 1972, "*Prediction of Effects of Mass-Transfer Cooling on the Blade-Row Efficiencies of Turbine Airfoils,*" AIAA January 1972.

J.H. Horlock, 1973, *Axial Flow Turbines,* Robert E. Keieger Publishing Company.

R.C. Keogh, G.R. Guenette, and T.P. Sommer, 2000, "*Aerodynamic Performance Measurements of a Fully-Scaled Turbine in a Short Duration Facility,*" ASME IGTI 2000-GT-486.

R.C. Keogh, 2001, "*Aerodynamic Performance Measurements of a Film Cooled Turbine Stage,*" Ph.D Thesis, Department of Aeronautics and Astronautics, MIT.

NIST14, 1992, " *NIST Mixture Property Database,*" NIST Thermophysics Division 838.02, Boulder, Colorado.

O.P. Sharma and T.L. Butler, 1987, "Predictions of Endwall Losses and Secondary Flows in Axial Flow Turbine Cascades," ASME *J. of Turbomachinary.* Also in *ASME Paper 86-GT-228.*

C.M. Spadaccini, 1999, "*Aerodynamic Performance Measurements of a Fully Scaled Film-Cooled Turbine Stage,*" Master's Thesis, Department of Aeronautics and Astronautics, MIT.

5 ROLE OF TIP CLEARANCE FLOW ON AXIAL COMPRESSOR STABILITY (E.M Greitzer, C. S. Tan)

5.1 Introduction and Background

Experiments performed on compressor rigs (Day, 1993, Longley et al., 1996, Silkowski, 1995, Tryfonidis et al., 1995, Weigl et al., 1997, Camp and Day, 1997) and engines (Day et al., 1997) have demonstrated that there are two routes to rotating stall in axial compressors. The first is characterized by the growth of small amplitude, essentially two-dimensional long wavelength disturbances (on the order of the circumference) which extend axially through the compressor. These disturbances, referred to as modal stall waves, can be detected from 10 to more than 100 rotor revolutions prior to stall, and they rotate at 0.2 to 0.5 times the rotor speed (McDougall, 1988, Haynes et al., 1996, Tryfonidis et al., 1995). Two-dimensional linearized stability analyses (Moore, 1984, Moore and Greitzer, 1986, Weigl et al., 1997) give a useful description of the modal wave shape, phase speed and growth rate for many compressors.

The second route to rotating stall is characterized by the development of three-dimensional "spike" disturbances (Day, 1993, Silkowski, 1995). These are localized to the tip region of a specific rotor in a multi-stage compressor and have a length scale on the order of the blade pitch. Spikes initially rotate at 0.7 to 0.8 times the rotor speed, but slow to 0.2 to 0.5 times the rotor speed as they develop into a full stall cell. The time between first detecting a spike and the final stall cell is less than five rotor revolutions, much shorter than for modal-type stall development.

Short wavelength stall inception behavior cannot be captured by linearized stability analyses. However more complex models such as that of Gong [1999], which incorporates a blade-row by blade-row body force representation of the compressor, enable simulation of aerodynamic instability inception local to a blade rotor tip and its development from a local flow event into the nonlinear limit cycle instability that involves the entire compressor (more specifically, Gong [1998, 1999] has successfully used such a flow model framework to simulate multi-stage compressor aerodynamic (linear as well as non-linear) instability inception and subsequent development into fully developed rotating stall). Thus it can be said that we understand and we know how to calculate the development from stall inception that involves localized three-dimensional spike disturbances into a fully developed rotating stall. What is not understood nor known is the connection between the formation of localized three-dimensional spike disturbance and the blade passage flow processes/events. In order words, the flow processes (at the blade passage level) responsible for providing the conditions for the instability mode (i.e. the localized spike disturbance) to grow is not understood; they are characterized by length scales of the order of blade-to-blade spacing or smaller. It is the lack of such understanding that motivated us to undertake the research reported under this task.

Isolated rotor calculations performed by Hoying et al. (1998) have implied a causal link between the rotor tip leakage flow and the formation of the incipient spike disturbances found in experiments. These together with the many experimental observations appear to implicate the role of rotor tip clearance flow on compressor instability onset, but the scenarios or the manner through which the tip flow evolves toward the formation of spike disturbance that grow into a rotating stall cell was not explained. Furthermore what set the experimentally observed length scale of the spike disturbance is not known nor explained. Thus the basic issue addressed in this task is the causal link between tip clearance flow and spike stall inception. Specifically, what is the mechanism for tip clearance flow to lead to the formation of short length-scale spike disturbances that grows into rotating stall cell and the conditions under which this occurs?

This section is organized as follows. The rotor passage flow feature common to compressors exhibiting spike stall inception is first identified. This allows subsequent focus on particular rotor geometry without foregoing the generic nature of the research. Time-accurate single blade passage computations are then used to identify a flow scenario and criteria setting single blade passage solution limit –or flow solution limit- (the lowest flow coefficient for which a physically consistent flow solution exists). This is followed by a six-blade passage flow simulation to clarify the link between tip clearance flow and spike disturbances, as well as the utility of single blade passage solution limit as guideline for predicting compressor stall inception. The findings are then used to explain the observed spike stall inception behavior.

5.2 A Common Rotor Computed Passage Flow Feature

Single blade passage computations (Vo, 2001) on different compressor rotor designs indicated that the flow solution limit is set by the interface between the incoming flow and tip clearance flow lining up with the rotor leading edge plane. These compressor designs are known to exhibit spike stall inception and thus this tip clearance flow behavior is hypothesized to be limiting the useful compressor flow range.

Figure 5.1 shows the trajectory of the interface between the incoming and tip clearance flows (Khalid *et al.*, 1999). As the flow coefficient decreases, the axial momentum of the incoming flow is reduced while the axial momentum of the tip clearance flow increases (from increased blade loading). The result is the movement of the interface toward the rotor leading edge plane, observed experimentally by Koch (1974). Since the tip clearance flow has higher entropy than the incoming flow, the interface can be identified as a region of high entropy gradient.

Contours of computed entropy on rotor blade tip radial plane at flow solution limit (based on steady-state Navier-Stokes computations) are shown in figures 5.2 and 5.3 for E^3 rotor B and NASA transonic rotor 35, respectively. The former is the first rotor of a four-stage low-speed (0.2 tip Mach number) axial compressor, which is known to exhibit spike stall inception (Silkowski, 1995) while the latter exhibited spike stall inception when subjected to inlet radial flow distortion (Spakovsky *et al.*, 1999).

These results show that a solution limit is reached when the incoming/tip clearance flow interface lines up with the rotor leading edge plane at the blade tip. Rotor casing oil flow pictures on a low-speed single stage axial compressor near the point of spike stall inception (Saathoff and Stark, 2000) also indicated such a limiting flow behavior.

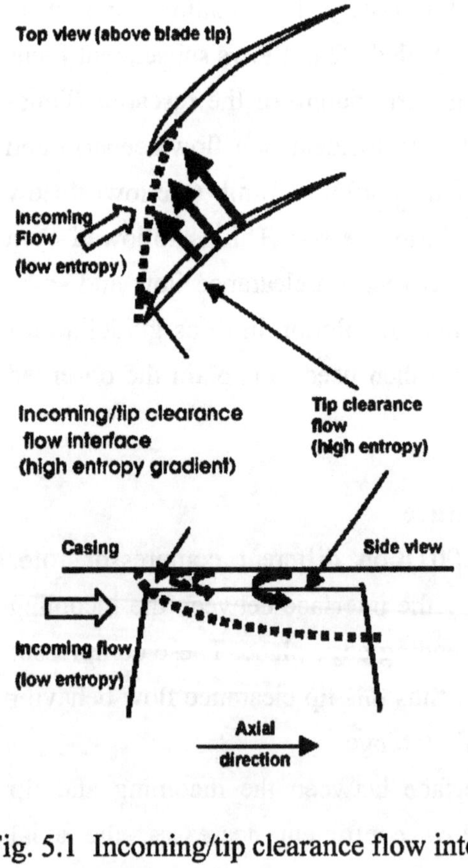

Fig. 5.1 Incoming/tip clearance flow interface

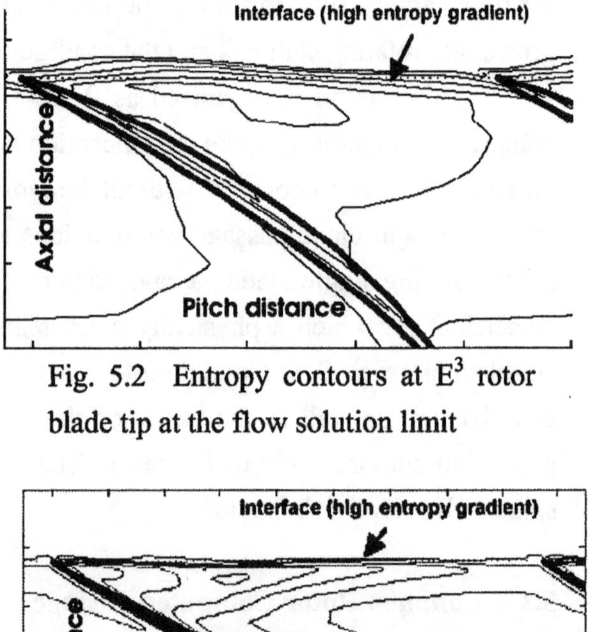

Fig. 5.2 Entropy contours at E^3 rotor blade tip at the flow solution limit

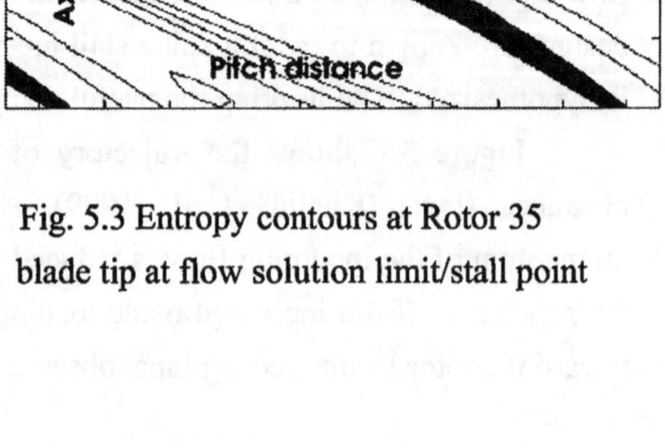

Fig. 5.3 Entropy contours at Rotor 35 blade tip at flow solution limit/stall point

The above results point to a role for the tip clearance flow in setting the flow solution limit and the formation of spike stall disturbances. The E^3 rotor B in figure 5.3 is selected for establishing the causal relationship between tip clearance flow and spike disturbances.

5.3 Passage Flow Solution Limit for E^3 Rotor B

Time-accurate single blade passage computations using UNSTREST (Denton, 1986) for E^3 rotor B are used to assess the behavior of the tip clearance flow beyond the flow solution limit, and to identify the threshold flow events leading to spike stall inception. This involves incrementally increasing the exit tip pressure to initiate the transient behavior that results in a rapid drop in mass flow beyond the flow solution limit. The computed total-to-static pressure rise characteristics depicted in figure 5.4 shows both the steady state flow regime (points 1 through 4) and transient regime (points 5 through 8) for a compressor rotor tip clearance size of 1.8% chord. The corresponding variation with respect to flow coefficient of total blockage and its individual contribution (tip clearance flow blockage and passage-blade/hub-boundary layer blockage) (Khalid *et al.*, 1999) are also indicated. The results show that the flow solution limit (point 4) marks the onset of rapid growth in blockage. This growth comes only from the tip blockage whereas the passage boundary layer blockage does not grow beyond the flow solution limit.

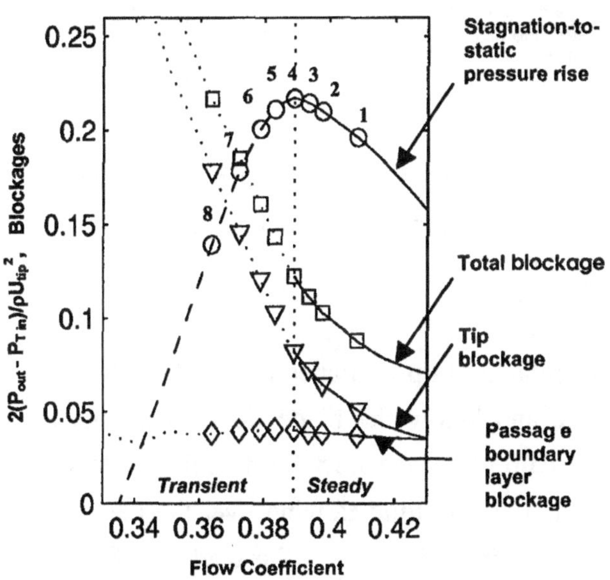

Fig. 5.4 Passage pressure rise and blockage variation for E^3 rotor B at 1.8% chord tip clearance

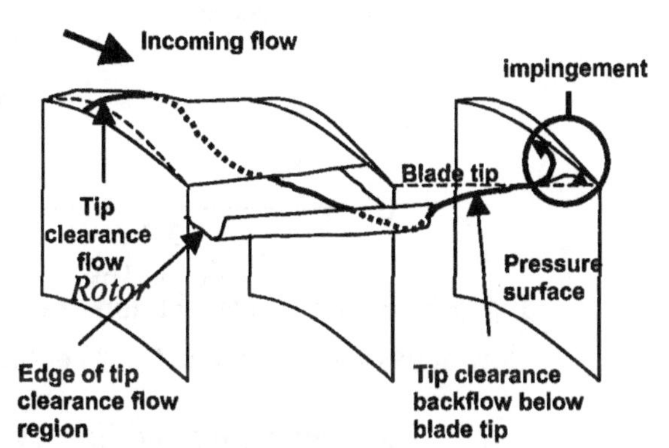

Fig. 5.5 Reversal ("backflow") of Tip clearance fluid below blade tip

Beyond this limit, the tip clearance flow from the adjacent blade passage is observed to flow into a blade passage at the trailing edge and impinge on the pressure surface as illustrated in figure 5.5. Henceforth, this reversal of the tip clearance fluid from the adjacent blade passage is referred to as "tip clearance backflow". The computed spanwise distribution of pitch-averaged mass flow in figure 5.6 shows the onset of tip clearance backflow below the blade tip at the flow solution limit. The operating point numbers correspond to those on figures 5.4. The mass flow per unit pitch is non-dimensionalized by inlet stagnation flow density times blade tip speed. For steady state solutions (points 1 through 3), figure 5.6 shows downstream (positive) mass flow below the blade tip. However, at the flow solution limit (point 4), the net mass flow reaches zero at the blade tip and becomes negative for transient points at lower flow coefficients (points 5 through 8), which indicates backflow below the blade tip.

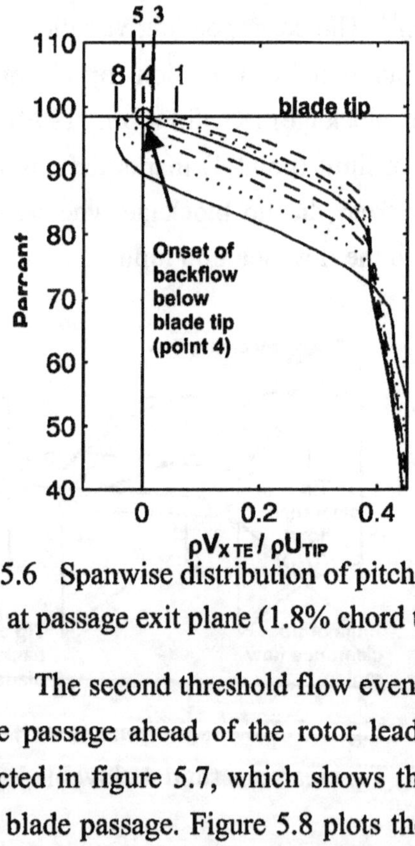

Fig. 5.6 Spanwise distribution of pitch-averaged mass flow at passage exit plane (1.8% chord tip clearance)

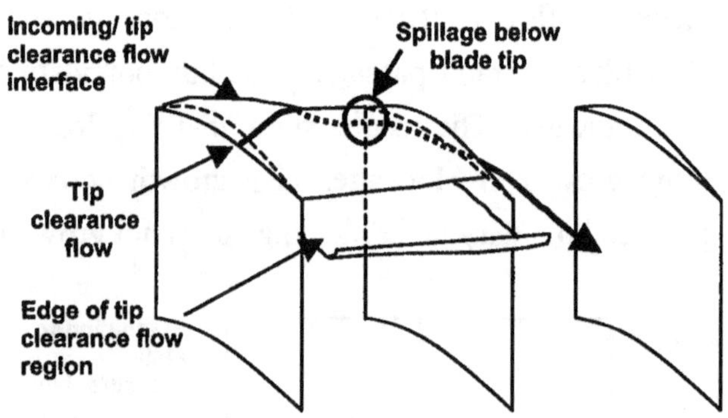

Fig. 5.7 Leading edge tip clearance Flow spillage below blade tip

The second threshold flow event is the spillage of tip clearance flow to the adjacent blade passage ahead of the rotor leading edge and below the blade tip radius. This is depicted in figure 5.7, which shows the path of tip clearance flow fluid spilling into the next blade passage. Figure 5.8 plots the mass flow of the fluid flowing over the blade tip and the mass flow spilling ahead of the blade leading edge below the blade tip for the operating points shown in figure 5.4. The curves, normalized by blade passage mass flow, show that the amount of spilled fluid increases while that of the fluid flowing over the blade tip decreases with decreasing flow coefficients below point 4. In particular, the bottom curve shows that the leading edge flow spillage below the blade tip starts to increase from near zero at the flow solution limit (point 4). The onset of tip clearance flow spillage below the blade tip means that the incoming/tip clearance flow interface crosses the leading edge blade tip. This is the common flow feature associated with compressor exhibiting spike stall inception identified in the previous section.

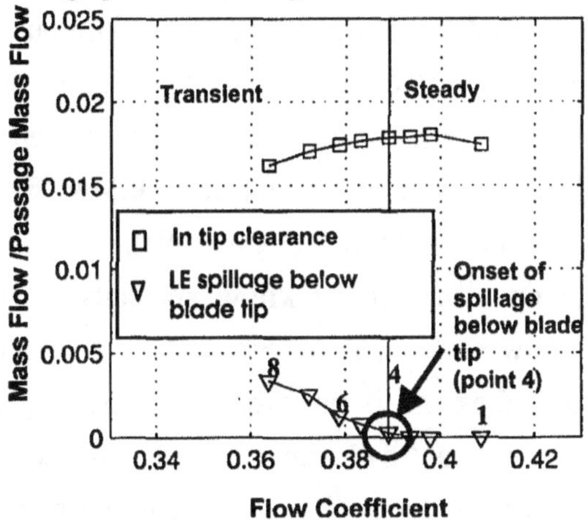

Fig. 5.8 Mass flow through blade tip and spilled ahead of leading edge below blade tip

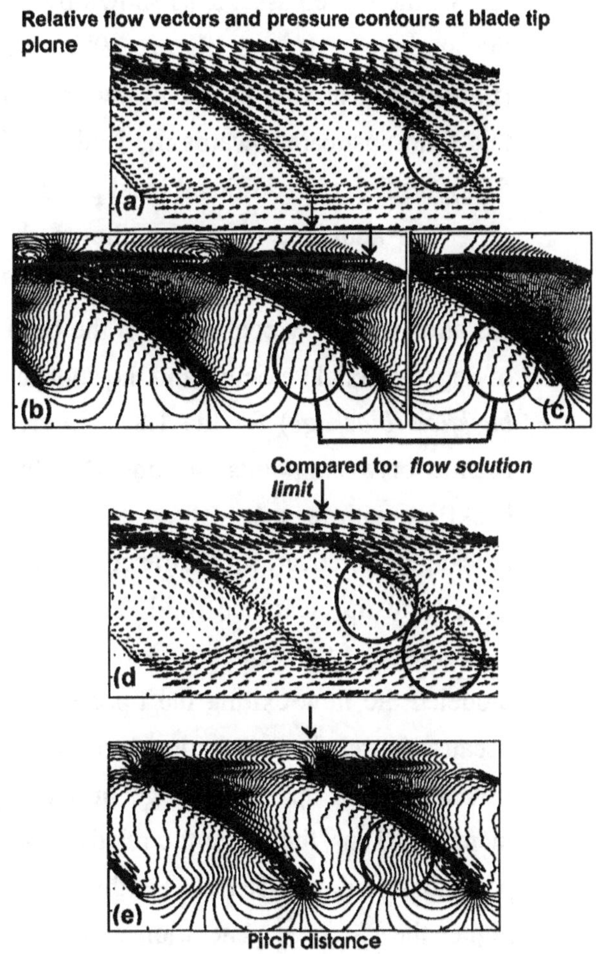

Fig. 5.9 Role of tip clearance backflow below the blade tip on flow solution limit

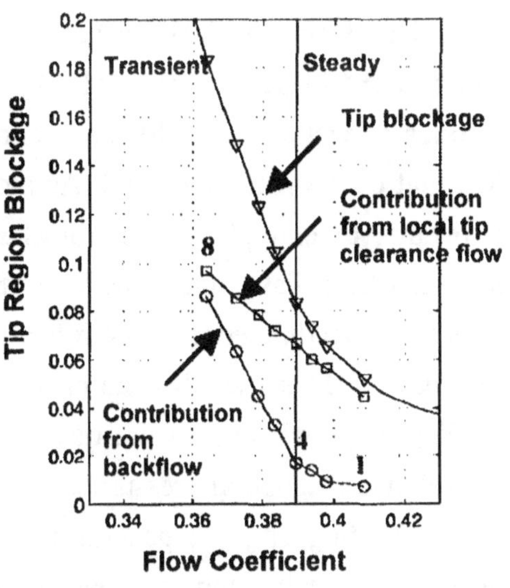

Fig. 5.10 Contribution of tip clearance backflow to tip blockage

We now proceed to explain how the above threshold flow events set the flow solution limit and contribute to the observed blockage growth. Figure 5.9 explains how the flow impingement elucidated in figure 5.5 leads to additional backflow and an increase of the tip clearance blockage. The flow features of interest are marked with circles. At the flow solution limit, there is no average backflow below the blade tip at the trailing edge as noted earlier; any increase in back pressure triggers backflow. The impingement of the backflow on the pressure surface (figure 5.9a) increases the local pressure gradient at the rear of the blade passage, as shown by the more closely-spaced static pressure contours in the impingement zone (figure 5.9b) when compared to those at the flow solution limit (figure 5.9c), driving the backflow upstream (figure 5.9d). By continuity, the movement of this fluid upstream towards the leading edge leads to an enhancement in backflow (figure 5.9d), resulting in additional local pressure rise from impingement (figure 5.9e).

In addition, it can be shown that the tip clearance backflow accounts for the observed increase in tip blockage beyond the flow solution limit. Figure 5.10 shows the variation with flow coefficient of the tip blockage, the contribution from backflow (regions of negative axial velocity) and the contribution from local tip clearance fluid (*tip blockage - backflow blockage*) for the 1.8% tip clearance case of figure 5.4. For the transient flow regime between points (4) and (8), the rate of increase in blockage associated with backflow with respect to flow coefficient accounts for most of the rate of increase in tip blockage. Moreover, tip clearance backflow is also responsible for the increase in blockage from local tip clearance fluid. This can be elucidated based on the blockage generation mechanism of tip clearance flow in compressors put forward by Khalid *et al.* (1999); he modeled the flow exiting the tip clearance gap as series of wakes along the blade chord because its relative stagnation pressure is lower than that of the incoming fluid. As these wakes convect up the passage pressure gradient towards to exit plane, their blockage increases. Thus, the lower the relative stagnation pressure of the wake and/or the larger the pressure rise it experiences (i.e. the nearer the wake is to the leading edge of the rotor) the higher the blockage generation will be. In this case, figure 5.11 shows that the backflow displaces the local tip clearance fluid toward the leading edge, where pressure rise to the passage exit plane is the greatest. This fluid has low relative stagnation pressure and as it leaks through the adjacent tip clearance gap near the leading edge, it generates more blockage in the next blade passage than if it were distributed equally along the blade chord (in the absence of backflow), in which case it would experience a lower pressure rise as it convects downstream in the adjacent blade passage.

The steps leading to the second threshold flow event is delineated in what follows. Figure 5.12 displays the relative stagnation pressure deficiency (with respect to the incoming flow) of the gap tip clearance fluid versus that of the tip clearance spillage fluid at point (6) in figure 5.8, implying that the tip clearance spillage fluid has lower relative stagnation pressure loss than if it had to leak through the adjacent tip clearance gap. Figure

5.13a shows that when the incoming/tip clearance flow interface is inside the blade passage (i.e. no leading edge spillage) the tip clearance fluid and backflow fluid can only convect downstream or leak through the adjacent tip clearance gap. However, as soon as leading edge spillage below the blade tip occurs (figure 5.13b), the tip clearance fluid precipitates toward the path of lower flow resistance allowing the backflow to propagate upstream.

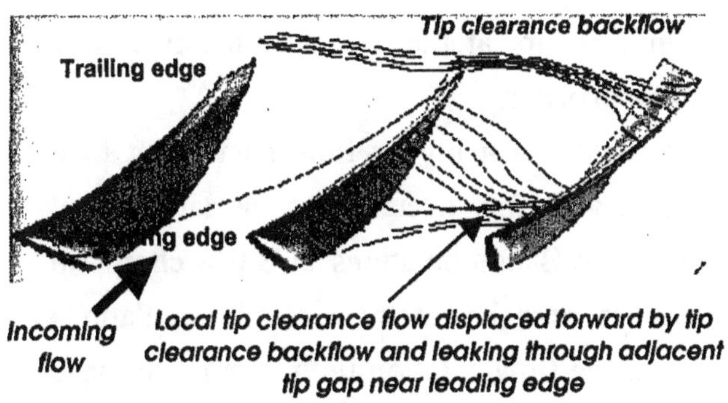

Fig. 5.11 Indirect contribution of tip clearance backflow to tip blockage.

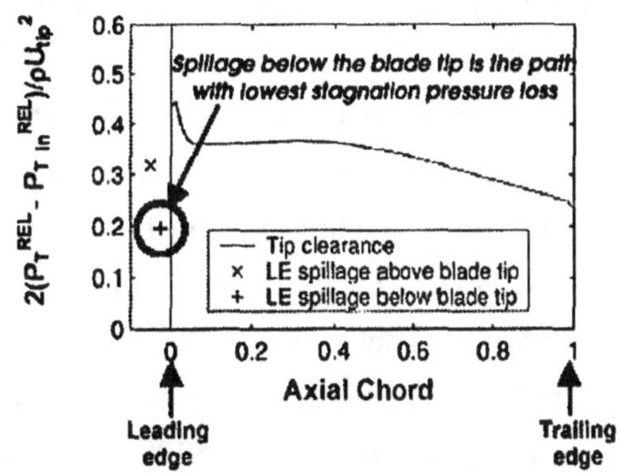

Fig. 5.12. Relative stagnation pressure deficiency (loss) of fluid flowing through the tip clearance gap and fluid spilled ahead of the leading edge for point 6 in fig. 5.8

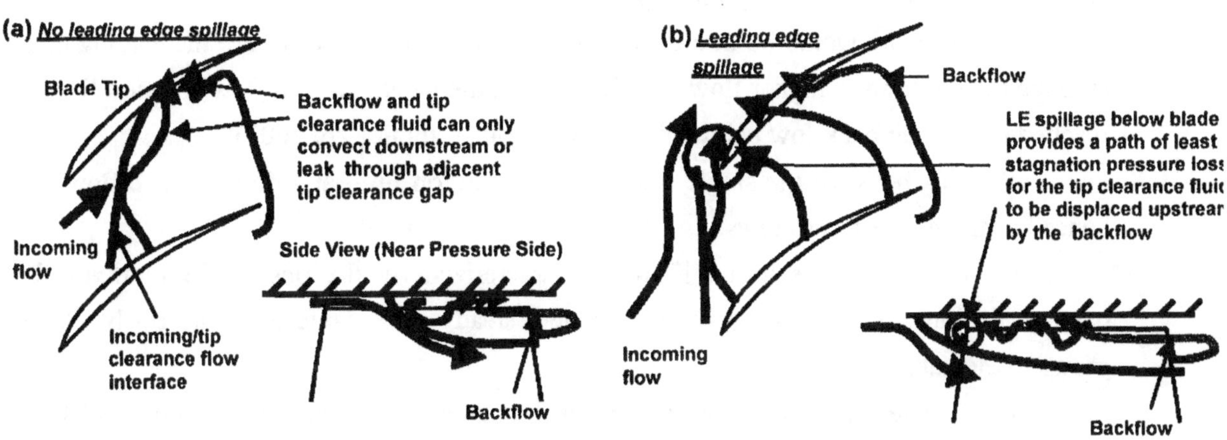

Fig. 5.13. Role of leading edge spillage below blade tip on upstream propagation of backflow

We deduced from the above the two criteria that set the blade passage flow solution limit as:

(1) Zero pitch-averaged mass flow across the pitch at the trailing edge blade tip (onset of tip clearance backflow)

(2) Onset of leading edge tip clearance flow spillage below the blade tip

Although the threshold flow events may start at different flow coefficients, both must be present (i.e. both criteria must be satisfied) for the flow solution limit to be reached. In other words, the flow solution limit occurs at the lower of the two flow coefficients where each threshold flow event starts to occur.

For example, tip clearance backflow can start at a higher flow coefficient than leading edge spillage, but the flow solution limit will only occur when the latter event comes about. This is demonstrated by the E^3 rotor B computations at 3.0% chord tip clearance. It is noted that for large tip clearance sizes, as is the case here, steady oscillations in the tip clearance flow were found to occur near the flow solution limit, similar to those observed by Mailach *et al.* (2000) and Marz *et al.* (2001). The time-averaged speedline, pitch-averaged mass flow and spillage mass flow distributions for this case are presented in figures 5.14a, b and c, respectively. The symbolic representations are the same as those in figure 5.4, 5.6 and 5.8, respectively. Figure 5.14b shows that backflow below the blade tip starts at point (1) but the flow solution limit does not occur until the onset of leading edge spillage below the blade tip at point 4 (figure 5.14c).

The opposite would be true for the small tip clearance case, where the leading edge spillage may occur at a higher flow coefficient than the backflow, but flow solution limit is reached only when backflow below the blade tip comes about (Vo, 2001).

The link between the single passage flow solution limit and tip clearance flow can hence be summarized as follows:

- The flow solution limit is characterized by the onset of growth of the passage tip blockage without settling to a new steady-state value until the tip blockage grows to the hub.
- The onset of this growth in tip blockage is marked by two threshold flow events that can start at different flow coefficients but must both be present:

(1) The backflow of tip clearance fluid from adjacent blade passages at the trailing edge plane below the blade tip, which impinges in the blade pressure side

(2) The spillage of tip clearance fluid ahead of the blade leading edge below the blade tip into the next passage.

The next section will connect the above description of the blade passage flow solution limit with the formation of spike disturbances. This link can be predicted based on four factors:

(1) The phenomena described occur in the tip region, where spike disturbances have been experimentally detected.

(2) The phenomena rely on the spillage of tip clearance fluid to the adjacent blade passage ahead of the rotor leading edge. This is consistent with the generic flow feature associated with spike stall inception where the incoming/tip clearance flow interface lines up with the rotor leading edge plane.

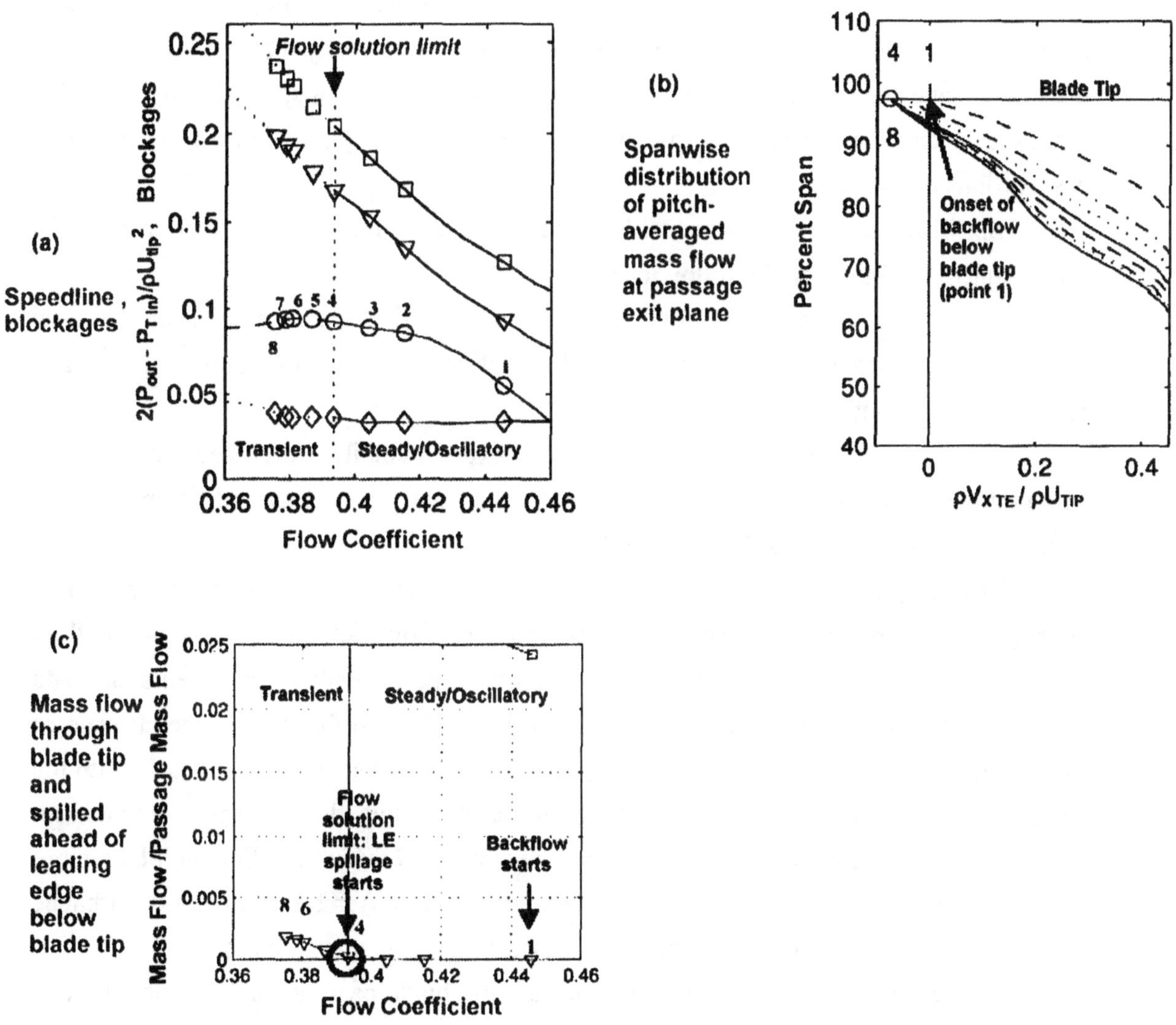

Fig. 5.14. Evaluation of passage flow solution limit criteria for 3.0% chord tip clearance

(3) We have observed the impingement pattern of trailing edge backflow in the stall computations by Hah *et al.* (1999) on another low-speed rotor.

(4) The required presence of both threshold flow events for stall inception can be observed in the study of rotating instabilities by Marz *et al.* (2001). Trailing edge backflow can been seen in their multiple blade passage computations in the stable flow range, but they claim that stall did not occur until the tip clearance flow leaks across the blade leading edge.

5.4 Link to Stall Inception

To link the proposed criteria to stall inception, a multiple blade passage computation beyond the flow solution limit was performed to verify that the criteria that set the single blade passage flow solution limit lead to the formation of spike disturbances. The proposed mechanism for spike formation was then assessed against known experimental observations on stall inception.

Six blade passages were considered sufficient to simulate the formation of spike stall disturbances based on the followings:

- The slope of the total-to-static pressure rise characteristics at the flow solution limit is negative so that modal stall inception is excluded (Moore and Greitzer, 1986; Camp and Day, 1998)
- Spike disturbances have been experimentally measured to have a circumferential extent of two to three blade passages (Day, 1993; Day and Freeman 1994; Dobat and Saathoff, 2001; Silkowski, 1995)

The computational code, mesh and procedure used were the same as for the single blade passage computations. The single blade passage flow field corresponding to flow solution limit was taken as the initial flow condition. The equivalence between single and multiple blade passage flow solution limit is evidenced by the fact that in the absence of any introduced asymmetry in the flow field, all blade passages stall simultaneously with the same flow transient beyond the flow solution limit as the single blade passage computation. Circumferential asymmetry is introduced into the flow field during the first ten blade-pitch convection times (well before any spike disturbance formation) in the form of a disturbance six blade passages in wavelength over the full span to avoid favoring the tip.

Figure 5.15 shows the results for the multiple blade passage stall simulation. The first plot displays the time history of the mass flow. The latter plots show the deviation from the passage-averaged axial velocity distribution, normalized by the blade tip velocity, at different times (marked in the mass flow time history) during the stall transient. The axial position chosen for the plots is 18% chord upstream of the rotor leading edge. This is representative of hot-wire locations used to measure spike disturbances (Day, 1993; Silkowski, 1995).

At time instant (1), when the initially imposed inlet stagnation pressure disturbance is removed, no visible velocity defect at the blade tip is seen, showing that the introduced flow asymmetry did not bias the results in favor of spikes. A short length-scale disturbance (in the form of an axial velocity deviation near the casing that is more pronounced at the crest than the peak of the wave) appears at time (3), accompanied by a rapid drop in the mass flow. A velocity defect near the casing and extending over two to three blade pitches becomes visible. The disturbance grows rapidly in amplitude and spreads toward the hub, as seen at time (4). It becomes more two-dimensional at time (5), covering the fundamental harmonic of the six-blade domain.

On the axial velocity deviation plots, the blades rotate from left to right along the *theta/pitch* axis. The velocity defect (short length-scale disturbance) rotates in the reverse direction (with respect to the blades) since it rotates slower than the rotor in the absolute frame of reference. The location of the peak of the short length-scale disturbance at different time steps allows estimation of the speed at which the disturbance rotates. The peak of the velocity defect has moved by about 3 blade pitches between time steps (3) and (4), which are 10 blade passing times (*BPT*) apart, and the speed of the disturbance is thus about 70% of the rotor speed. The time to grow into fully developed stall can be estimated by the time it takes for the disturbance from inception (~time (3)) to extend across the entire span (~time (5)), which is about two rotor revolutions.

Table 5.1 compares the computed short length-scale disturbance to the experimentally measured spike disturbance in front of the E^3 rotor B obtained from hot-wire traces (Silkowski, 1995) as summarized by Gong *et al.* (1999). The disturbance obtained from the computations compares well with the measured features. However, because the length-scale of the computed disturbance (two-to-three blade passages) covers about half of the fundamental wavelength of the domain (six blade passages), computations with more blade passages should be pursued to ensure that the number of passages used in the computation have no influence on the resulting disturbance.

The next step is to verify that the scenario for the formation of the spike disturbance is consistent with that behind the single blade passage flow solution limit. This is done by examining the tip clearance flow behavior, in terms of tip clearance backflow at the trailing edge and leading edge tip clearance fluid spillage, for the blade passage behind the spike disturbance.

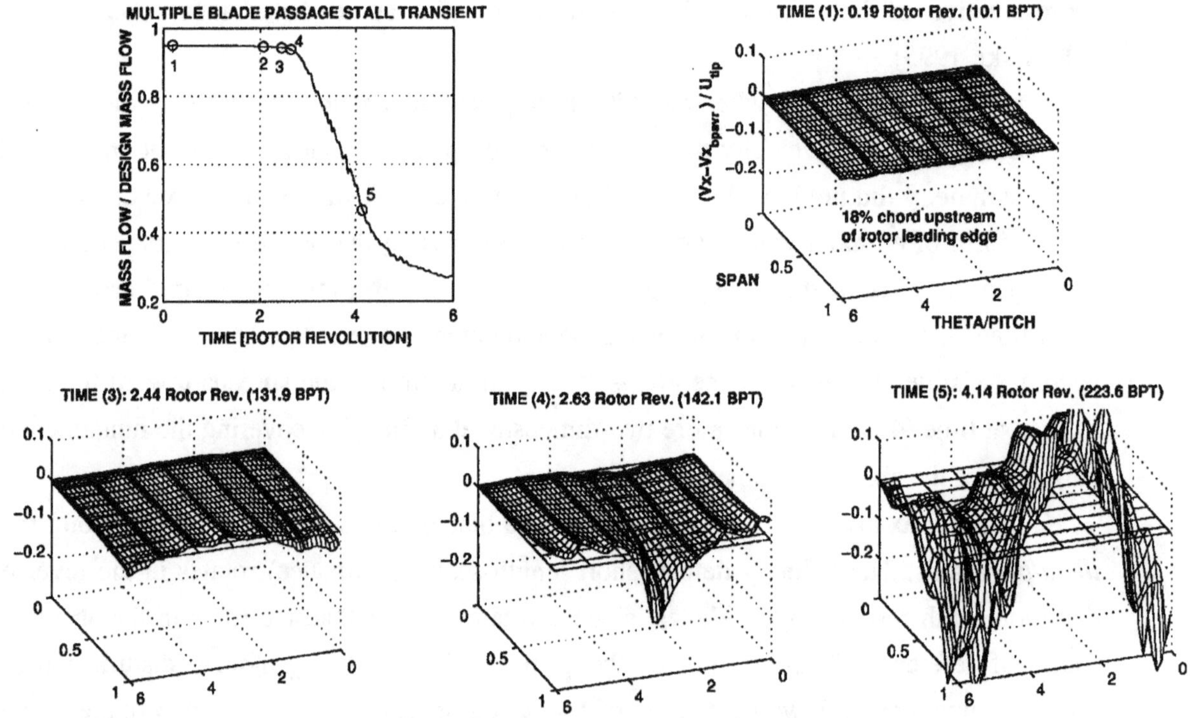

Fig. 5.15 Stall transient for multiple (6) blade passage configuration

Table 5.1. Comparison of measured (Gong et al., 1999) and computed spike disturbance

Characteristics	Experiment	Computation
Circumferential Extent	~2-3 blade pitch	~2-3 blade pitch
Radial Extent	concentrated in tip region	last 10-20% span
Rotating Speed	70-73% of rotor speed	~70% of rotor speed
Growth Time to Full Stall Cell	~3 rotor revolutions	~2 rotor revolutions

Figure 5.16 displays the relative flow vectors at the blade tip radius at time (4). At this time the velocity defect (spike disturbance) is in front of blade passages 3 and 4 and is moving towards blade passage 5. Figure 5.17 shows the corresponding spanwise distribution of pitch- averaged mass flow and leading edge tip clearance fluid spillage below the blade tip, for each blade passage. From figures 5.16 and 5.17a, passages 3 and 4 show trailing edge backflow and impingement below the blade tip due to tip clearance fluid

from adjacent blade passages. The other blade passages do not exhibit any trailing edge backflow below the blade tip.

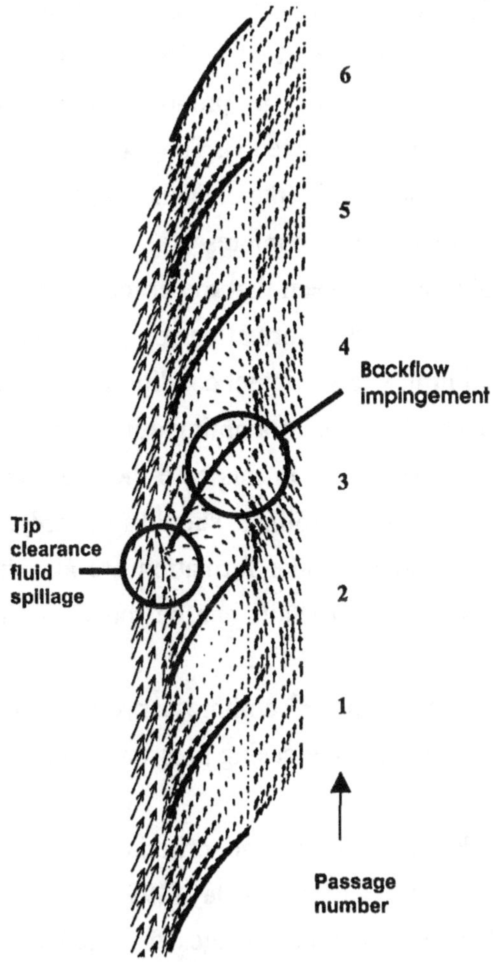

Fig. 5.16 Relative flow vectors at blade tip at time 4.

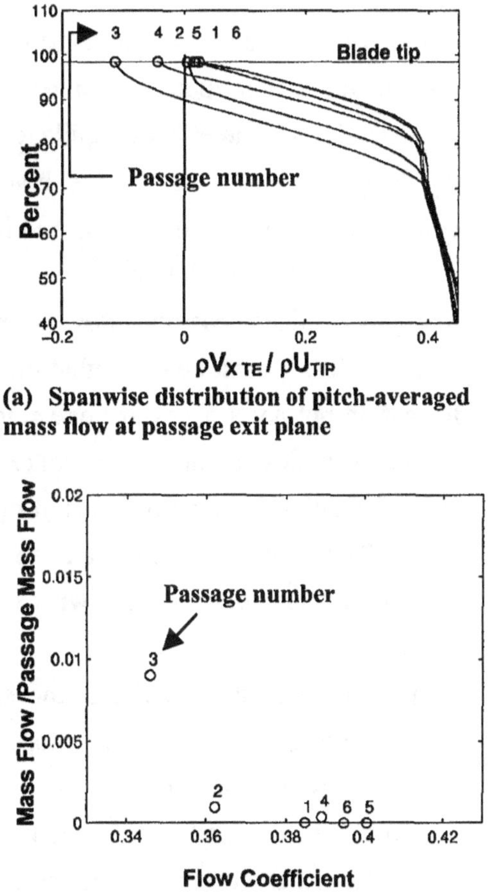

(a) Spanwise distribution of pitch-averaged mass flow at passage exit plane

(b) Leading edge spillage of tip clearance fluid below the blade tip

Fig. 5.17 Tip clearance flow behavior in each blade

Figures 5.16 and 5.17b also show leading edge spillage of tip clearance fluid below the blade tip from blade passage (3) into passage (4). These two flow events are similar to those observed in the single blade passage flow transient past the flow solution limit, for which the blade-to-blade periodic condition is no longer valid in the computed multiple blade passage flow. However, it is to be noted that under the conditions that the computational experiments have been carried out, the flow computed in a single blade passage and in the multiple blade passage is identical up to the flow solution limit.

Moreover, the two flow events set the experimentally observed spike disturbance length-scale of two-to-three blade pitches. As shown in figure 5.16, a minimum of three blade passages are involved in the spike formation. All the leading edge spillage fluid from the passage (3) flows into passage (4), such that the disturbance is most pronounced for these two passages. The backflow tip clearance fluid may come mainly from passage (2), but also from one or two passages beyond. This means that the effect of backflow on the disturbance is distributed among several passages and thus not as pronounced for the passage (2) as is the effect of spillage on the disturbance for passage (3). This explains why the length-scale is two-to-three blade pitches and not simply three. This length-scale concurs with that of spikes measured in experiments on different compressors.

From the multiple blade passage stall simulation, it is seen that the criteria that set the single blade passage flow solution limit lead to the formation of spike stall disturbances. It must be noted that the flow solution limit is a sufficient condition for spike formation but not a necessary one since a disturbance of sufficient amplitude introduced a higher mass flow could trigger spike formation. However, even in this case the proposed spike formation scenario in terms of tip clearance backflow and spillage should still apply.

The computed results and the proposed scenario for spike formation are also in accord with the following known experimental observations (details can be found in Vo (2001)):

(1) the best place to detect the spike disturbance on the E^3 compressor was at the trailing edge of the first rotor near the casing, using either casing pressure sensor or hot-wires (Park ,1994);
(2) the blade loading at which both the backflow and spillage criteria for spike formation corresponds to the critical tip incidence of Camp and Day (1998) ;
(3) as the tip clearance is increased, the stall inception type changes from spike to modal, as observed by Day (1993).

The results from this paper can be used to explain the observations by Lee and Greitzer (1990). Lee and Greitzer (1990) measured the effect of endwall suction and blowing on compressor stability, using a cantileverd stator and rotating hub to simulate tip clearance flow. Because their work occurred prior to the discovery of either modal or spike stall inception, no information of the stall inception type can be obtained. However, their main concern with endwall stall combined with the negative slope of the static pressure rise characteristic at stall (in the presence of endwall slots) point to spike stall inception. They reported that rear endwall suction and/or leading edge blowing, as would be associated with casing treatment, were effective in increasing stall margin. According to the proposed mechanism for spike formation, trailing edge suction of the tip clearance backflow and leading edge blowing, which would tend to delay the movement of the incoming/tip clearance flow interface to the leading edge plane, should be effective in delaying the formation of spike disturbances, and thus spike stall inception. The view here is that any technique that delays either of the two criteria that set passage flow solution limit should be effective in delaying spike stall inception.

5.5 Conclusions

In summary the conclusions are as follows:

(1) The growth in tip blockage at the single blade passage flow solution limit is caused by the impingement on the trailing edge pressure surface of backflow from tip clearance fluid of adjacent blade passages, and by leading edge spillage of tip clearance flow below the blade tip to the adjacent passage.

(2) Zero mass flow across the pitch at the trailing edge blade tip, and the onset of leading edge tip clearance flow spillage below the blade tip (both evaluated on a time-averaged basis) form two criteria that set the lowest flow coefficient at which a single rotor blade passage can operate in equilibrium. They also signal the onset of spike disturbances.

(3) The proposed scenario for the formation of spike disturbances is consistent with published observations on stall inception behavior of axial compressors, including the observed length-scale of spike disturbances, thus reinforcing its plausibility.

The implications from these are:

(1) For compressors that exhibit short length-scale stall inception, a flow alteration methodology that delays one of the two proposed criteria for spike formation should be successful in extending the mass flow operating range.

(2) The present work suggests that single blade passage computations may be used to predict stall inception if they have the following attributes:

- Exit flow boundary conditions that can capture steady-state solutions past the peak pressure rise of the rotor blade passage

- Time-accurate computations so that time-averaged flow field from oscillatory equilibrium solutions can be used to evaluate the criteria for spike formation. Steady-state computations are thus not suitable for cases where unsteadiness from tip clearance flow oscillations or adjacent blade rows are present.

- Single blade passage solution for multiple blade rows to obtain the stagnation-to-static pressure rise characteristics of each blade row and of the entire compressor. This can be used to predict modal stall inception (Moore and Greitzer, 1986) and growth of spikes disturbances to fully developed rotating stall (Gong *et al.*, 1999).

5.6 References

Camp, T.R. and Day, I.J., 1998, "A Study of Spike and Modal Stall Phenomena in a Low-Speed Axial Compressor", *ASME Journal of Turbomachinery*, Vol. 120, pp.393-401.

Chen, G.T., Greitzer, E.M., Tan, C.S. and Marble, F.E., 1991, "Similarity Analysis of Compressor Tip Clearance Flow Structure", *ASME Journal of Turbomachinery*, Vol. 113, pp.260-271.

Day, I.J., 1993, "Stall Inception in Axial Flow Compressors", *ASME Journal of Turbomachinery*, Vol. 115, pp.1-9.

Day, I.J. and Freeman, C., 1994, "The Unstable Behavior of Low and High Speed Compressors", *ASME Journal of Turbomachinery*, Vol. 116, pp.194-201.

Denton, J.D., 1986, "The Use of a Distributed Body Force to Simulate Viscous Effects in 3D Flow Calculations", ASME Paper No. 86-GT-144.

Dobat, A. and Saathoff, H., 2001, "Experimentelle Untersuchungen zur Entstehung von Rotating Stall in Axialventilatoren", VDI-Berichte.

Furukawa, M., Inoue, M., Saiki, K. and Yamada, K., 1999, "The Role of Tip Leakage Vortex Breakdown in Compressor Rotor Aerodynamics", *ASME Journal of Turbomachinery*, Vol. 121, pp.469-480.

Gong, Y., Tan, C.S., Gordon, K.A. and Greitzer, E.M., 1999, "A Computational Model for Short Wavelength Stall Inception and Development in Multi-Stage Compressors", *ASME Journal of Turbomachinery*, Vol. 121, pp.726-734.

Hah, C., Schulze, R., Wagner, S., Hennecke, D.K., 1999, "Numerical and Experimental Study for Short Wavelength Stall Inception in a Low-Speed Axial Compressor", Paper AIAA 99-IS-033.

Khalid, S.A., Khalsa, A.S., Waitz, I.A., Tan, C.S., Greitzer, E.M., Cumpsty, N.A., Adamczyk, J.J. and Marble, F.E., 1999, "Endwall Blockage in Axial Compressors", *ASME Journal of Turbomachinery*, Vol. 121, pp.499-509.

Koch , C.C., 1974, "Discussion of Benser WA: Transonic Compressor Technology Advancements", *Fluid Mechanics, Acoustics, and Design of Turbomachinery*, Part II, NASA SP-304.

Lee, N.K.W. and Greitzer, E.M., 1990, "Effects of Endwall Suction and Blowing on Axial Compressor Stability Enhancement", *ASME Journal of Turbomachinery*, Vol. 112, pp.133-144.

Mailach, R., Lehmann, I. and Vogeler, K., 2000, "Rotating Instabilities in an Axial Compressor Originating From The Fluctuating Blade Tip Vortex", ASME Paper 2000-GT-506.

Marz, J., Hah, C. and Neise, W., 2001,"An Experimental and Numerical Investigation Into the Mechanisms of Rotating Instability", ASME Paper 2001-GT-536.

Moore, F.K., and Greitzer, E.M., 1986, "A Theory of Post-Stall Transients in Axial Compressors, Part I - Development of Equations", *ASME Journal of Engineering for Gas Turbines and Power*, Vol. 108, pp.68-76.

Park, H.G., 1994, "Unsteady Disturbance Structures in Axial Flow Compressor Stall Inception", M.S. Thesis, MIT Department of Aeronautics and Astronautics.

Saathoff , H. and Stark, U., 2000, "Tip clearance Flow Induced Endwall Boundary Layer Separation in a Single-Stage Axial-Flow Low-Speed Compressor", ASME Paper 2000-GT-0501.

Silkowski, P.D., 1995, "Measurements of Rotor Stalling in a Matched and Mismatched Multistage Compressor", GTL report No. 221, Massachusetts Institute of Technology.

Spakovszky, Z.S., Weigl, H.J., Paduano, J.D., Van Schalkwyk, C.M., Suder, K.L. and Bright, M.M., 1999, "Rotating Stall Control in a High-Speed Stage with Inlet Distortion: Part I- Radial Distortion", *ASME Journal of Turbomachinery*, Vol. 121, pp.510-516.

Storer, J.A. and Cumpsty, N.A., 1991, "Tip Leakage Flow in Axial Compressors", *ASME Journal of Turbomachinery*, Vol. 113, pp.252-259.

Vo, H.D., 2001," Role of Tip Clearance Flow on Axial Compressor Stability", Ph.D. Thesis, MIT Department of Aeronautics and Astronautics.

6.0 IMPACT OF TIP CLEARANCE FLOW ON CENTRIFUGAL IMPELLER PUMP PERFORMANCE **(E. Greitzer, C. S. Tan)**

6.1.0 Introduction

6.1.1 Background

Design trends for rocket engines are leading to the use of unshrouded impellers for the turbopumps. Removing the shroud reduces blade stresses and makes possible an increase of rotor tip speed, a decrease in diameter, and a decrease in weight.

The price paid for this weight reduction, however, is tip clearance during operation. Tip clearance has detrimental effects on pump performance[1], affecting both pressure rise capability and efficiency. It is therefore important to identify the parametric dependence of the performance sensitivity in order to provide guidelines for pumps, which are less sensitive to operational tip clearance change.

Although there has been much research on tip clearance effects in axial machines, there is less information on centrifugal machines. Further while a number of computational and experimental studies (see Section 6.1.2) have been conducted, there appears to be no readily accessible physical model of the situation. This Task addresses this lack and examines tip clearance effects in low specific speed centrifugal impellers.

6.1.2.0 Literature Survey

6.1.2.1 Sensitivity Trends and Correlations

The performance of a centrifugal pump can be characterized in terms of the static pressure rise coefficient CP_S, the stagnation pressure rise coefficient CP_T and the efficiency η. The expressions defining these measures are respectively:

- Static pressure rise coefficient, $CP_S = \dfrac{\overline{P_S}^M - \overline{P_{S,inlet}}^M}{\rho U_{tip}^2}$
- Stagnation pressure rise coefficient, $CP_T = \dfrac{\overline{P_T}^M - \overline{P_{T,inlet}}^M}{\rho U_{tip}^2}$
- Efficiency, $\eta = \dfrac{CP_T}{W_{input}}$.

The operator $\overline{\quad}^M$ means mass average. We choose to use mass average because the flow is non-uniform in the impeller, and static and stagnation pressure rise across the impeller are derived from the Euler Turbine Equation, which is an energy flux equation.

---

[1] In this section, we use the word "performance" to denote both pressure rise and efficiency

A way to quantify the effects of tip clearance is by evaluating the change in static pressure rise coefficient CP_S, stagnation pressure rise coefficient CP_T and efficiency η with tip clearance ratio. We define a sensitivity parameter s_x as

$$s_X = -\frac{\Delta\left(\dfrac{X - X_0}{X_0}\right)}{\Delta\lambda},$$

where λ denotes the *tip clearance ratio* (tip clearance gap divided by impeller outlet blade span) , X denotes the performance measure (CP_S, CP_T or η) and the subscript 0 denotes the value corresponding to zero tip clearance (shrouded case, shroud rotating with the blade[2]). The sensitivity parameter s_x is expressed in % change of X per percent change of tip clearance ratio. If the change in X is proportional to λ (a common observation for tip clearance effects), s_x is a constant.

Table 6.1-1 sums-up the stagnation pressure rise and efficiency sensitivity to tip clearance variation, as measured or computed by different authors/investigators for centrifugal impellers. All published results that could be found on the subject are shown in this table.

Table 6.1-1 shows that efficiency sensitivity varies depending on pump design. There is a factor of almost four in the efficiency sensitivity reported by Brasz [1988] and Hoshide and Nelson [1972]. Ishida and Senoo [1985] also conducted performance tests using an impeller with three configurations of blade tip (round edge, sharp square edge and edge with an end-plate). The impeller response to tip clearance increase was observed to depend on blade tip configuration, resulting in the wide range of sensitivity for their impeller as reported in Table 6.1-1.

Several authors have tried to isolate which parameters are critical for the performance sensitivity to tip clearance. Engeda [1987], [1988], and [1995] concluded that the specific speed was not the only parameter that characterized tip clearance effects

---

[2] It was decided to use the shrouded case and not the 0% tip clearance case (with stationary casing) as a reference in the sensitivity definition. This choice is motivated by the fact that the shrouded case corresponds to a real situation, while 0% tip clearance is an artificial case. As will be seen in Section 6.4.1.2, the difference between pressure rise in shrouded and 0% tip clearance cases is twelve times less than the difference between pressure rise in shrouded and 4% tip clearance cases. As a consequence, the choice of shrouded or 0% tip clearance case as a reference has a negligible effect on the sensitivity value.

and this single parameter did not capture the overall behavior. Lauer et al. [1997][3] published experimental data based on 14 unshrouded impellers with varying tip clearance. A correlation parameter based on impeller geometry was proposed to characterize impeller pressure rise sensitivity to tip clearance variation, but the authors concluded that calculations on further pumps would be required to prove the validity of the correlation parameter. The results reported in [Senoo, 1991] showed that the blade number and exit angle were the main governing parameters for tip clearance impact on impeller performance and that pump performance is more sensitive when you increase blade exit angle or decrease the number of blades.

		Design Characteristics					Performance Sensitivity	
	Investigation Method	Specific Speed	Flow Coefficient	Number of Blades	Backsweep Angle (deg, from radial)		Efficiency Sensitivity	Stagnation Pressure Rise Sensitivity
Hoshide and Nelson (1972) [2]	Experiments	0.79	0.11	5 +5 splitter	32		1.20	1.40
Brasz (1988) [3]	Experiments	0.41			30		0.31	0.65
Ishida, Senoo (1981-1990) [4,5,6]	Experiments	0.43	0.27	16	45	Baseline Configuration	0.80	1.00
						Round Edge	0.60	0.94
						Sharp Square Edge	0.55	0.83
						Edge with an End-Plate	0.40	0.57
Goto (1992) [7]	Experiments/ Computations	1.34	0.17	5	64			2.50
Williams et al. (2001) [8]	Experiments/ Computations	0.23	0.12	6 +6 splitter	41			0.79-1.14

Table 6.1-1: Sensitivity of Different Impellers to Tip clearance

6.1.2.2 Existing Models

At present, there appears to be no consistent flow model of tip clearance effects. The closest is the results of Senoo, [1985] and [1991]. Although he conducted an in-depth

[3] The articles by Engeda, etal.[1987, 1988, 1995] and Lauer, etal.[1997] do not provide detailed results concerning the performance sensitivity of the different studied impellers, so we could not report results for these studies in table 6.1-1.

analysis, his results are based on an aerodynamic analogy (using drag and induced drag generated by the tip leakage flow). As such they are difficult to translate into turbomachinery parameters (for example flow turning).

In published work, mixing of the tip leakage flow entering the channel with the surrounding flow is treated as a local process, i.e. it takes place in a region whose extent from the blade tip is a few tip clearance gaps [Denton, 1993]. It is also commonly found that the major part of the mixing has occurred before the tip vortex roll-up location [Khalid, 1995].

Different models for the entropy generation due to tip clearance exist (for example the model of Storer-Cumpsty [Denton 1993] and a model based on the loss of cross-flow energy [Senoo, etal. 1985], [Senoo, 1995]). These models correspond to different level of mixing of the tip leakage flow with the core flow, namely partial [Senoo, etal., 1985] or complete [Denton 1993]. They are therefore appropriate in different situations. This distinction is not clear in the existing literature.

The next two subsections detail and compare these two models for mixing loss.

6.1.2.2.1 Loss of Cross Flow Energy Model

The tip leakage flow enters the channel with the same reduced stagnation pressure[4] as the core flow. However, its velocity $\mathbf{W_c}$[5] has a component W_n normal to the free stream velocity $\mathbf{W_m}$, and the magnitude of its streamwise component is lower than $|\mathbf{W_m}|$. The loss source is regarded as the loss of tip leakage cross-flow kinetic energy, as shown in Figure 6.1-1.

A justification of this description is provided by Senoo [1991], who assumes that the tip leakage flow generates a drag and derives an expression for this drag using an aerodynamic analogy. He identifies the loss associated with this drag to the loss of tip leakage cross-flow energy.

With this description of the mixing, we can identify the flux of generated entropy in the main stream to the flux of tip leakage cross-flow kinetic energy entering the channel. The entropy generation per unit mass in the main stream is thus

---

[4] $P_{T,red} = P_S + \frac{\rho W^2}{2} - \frac{\rho U^2}{2}$ with P_S the static pressure, W the relative velocity magnitude and U the local rotational speed of the impeller.

[5] Bold characters refer to vectors (see Nomenclature)

$$\frac{ds}{c_P} = \frac{1}{2}\frac{d\dot{m}_c W_n^{\ 2}}{\dot{m}_m c_P T}, \qquad (6.1\text{-}1)$$

where $\dot{m}_m$ is the mainstream flow rate, $d\dot{m}_c$ is the tip leakage flow rate, T is the temperature, s is the entropy and W_n is the component of tip leakage velocity normal to the core flow.

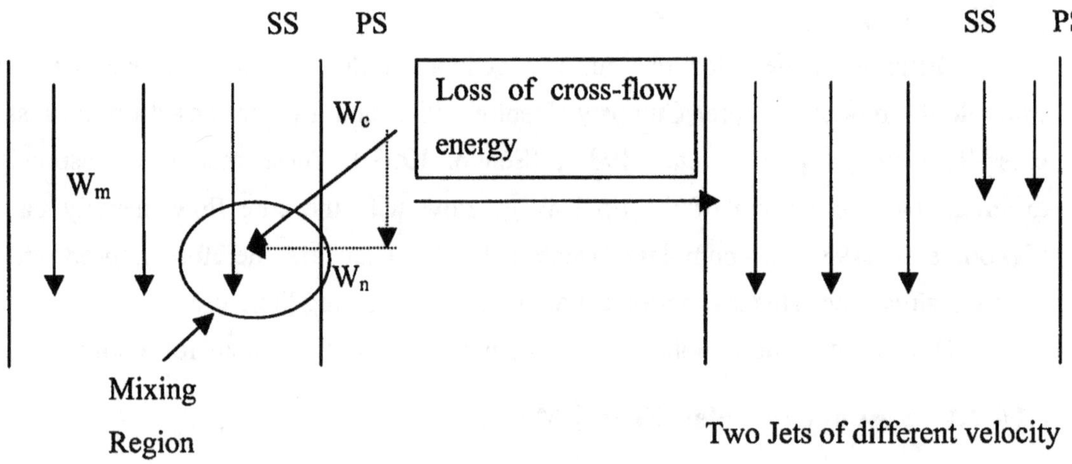

Figure 6.1-1: Loss of Cross Flow Energy Model

6.1.2.2.2 Storer-Cumpsty Model [Denton, 1993]

Denton [1993] describes the entropy generation associated with the injection of a secondary stream into a main stream (See Figure 6.1-2). This model can be applied to the tip clearance situation by discretizing the blade passage into small blade segments, each of them reproducing the injection situation. We consider that the rate of the leakage flow is much smaller than that of the main flow and carry out a control volume analysis to derive the entropy change in the main flow. Although, this model assumes that the mixing occurs at constant area, Denton [1993] states that the dependence on area has only a small effect on the overall result.

Applying the control volume shown in Figure 6.1-2 to conservation of momentum and continuity, we obtain

$$\dot{m}_m T ds = d\dot{m}_c \left(W_m^{\ 2} - W_m W_c \cos\alpha\right). \qquad (6.1\text{-}2)$$

The velocity magnitude of the main flow and the leakage flow are equal, and Equation 6.1-2 can be written as

$$\dot{m}_m T ds = \frac{1}{2} d\dot{m}_c (W_c \cos\alpha)^2 + \frac{1}{2} d\dot{m}_c (W_m - W_c \sin\alpha)^2 . \tag{6.1-3}$$

The first term on the right hand size corresponds to the loss of tip leakage cross flow kinetic energy, while the second corresponds to the loss kinetic energy defined based on the streamwise component of the leakage flow relative to the main stream flow.

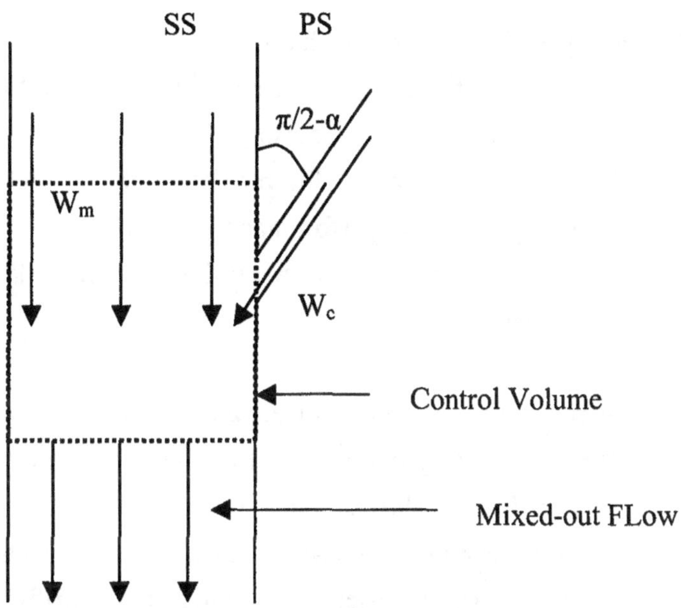

Figure 6.1-2: Model of Storer-Cumpsty: Control Volume

The difference between the two mixing models is the second term. With the loss of cross flow energy model, the flow is considered only partially mixed so a difference exists between the streamwise component of leakage and main-flow velocities. The model of Storer-Cumpsty assumes the tip leakage flow has completely mixed.

6.1.2.2.3 Blockage Models

The models described above only predict one part of the impeller performance decrease due to tip leakage flow. For centrifugal impellers, we have not found in the literature any simple model that accounts for the effects of blockage due to tip leakage flow.

For axial compressors, Khalid [1995] has developed a two dimensional blockage model to describe the evolution of the blocked area along the blade passage. The blade passage was discretized into neighboring discrete blade segments. The tip leakage flow over each segment generates a discrete wake that proceeds downstream in a pressure gradient. The discrete wakes are assumed to be independent and, adding their contributions, the blocked area at impeller exit can be found. For axial compressors, this model was in good agreement with experimental results. It will be seen below that Khalid's approach can be used also for centrifugal impellers (with a quasi-one-dimensional inviscid model for wake evolution in a pressure gradient [1999]).

6.1.3 Problem Statement

The primary goals of this research task were to:

- Define the differences between tip clearance flow phenomena in centrifugal pumps and in axial compressors.
- Develop a flow model for predicting the impact of tip clearance variation on centrifugal impeller performance
- Identify the parameters governing the impeller performance sensitivity to performance.

The following research questions were addressed:

- What physical processes control the formation and evolution of tip clearance vortices in an unshrouded impeller?
- Does a quasi-one-dimensional blockage model based on the loss of cross flow kinetic energy give useful information for centrifugal impellers?

6.1.4 Key Results

The key results accomplished in this Task are as follows:

- A simple flow model for the effect of tip clearance on impeller performance (pressure rise and efficiency) has been developed and assessed against CFD and published results. The model was based on the loss of cross flow kinetic energy and the evolution of the resulting wake-like flow in a pressure gradient (blockage), identifying the same mechanisms as described for axial compressors [Khalid, 1995]. This shows the generic nature of tip clearance phenomena for machines ranging from low specific speed (pumps) to high specific speed (axial compressors).
- Using the flow model, approximate expressions have been developed to describe the performance sensitivity to tip clearance variation and the parameters that affect performance sensitivity. From these expressions a set of guidelines for decreases in tip clearance sensitivity in centrifugal impellers can be deduced.

- A method was developed to visualize secondary flow structure on any reference plane. The tip vortex center trajectory in a centrifugal impeller was found to follow a similarity rule, analogous to that developed by Chen and etal. [1991] for axial compressors.

6.2.0 Technical Approach

6.2.1 Introduction

The approach to accomplish the technical objectives consisted of: (1) designing and implementing computational simulations to assess and explain the sensitivity to tip clearance of two centrifugal impellers; and (2) developing a simple flow model for capturing the performance sensitivity and identifying the controlling impeller parameters. These are elaborated below.

6.2.2 Impeller Choice

We have based our investigation on two impellers, one of which is representative of that for space applications. The impeller geometry was made available by Katz of Johns Hopkins University and Lee of the Naval Research Laboratory. The design characteristics of the impeller are summarized in Table 6.2-1. The specific speed is in the range for rocket pumps, but higher than Rocketdyne's prototype impeller (Ns=0.28) [Williams, etal. 1991].

Geometry		Design Parameters	
Number of Blades	5	Exit Flow Coeffcient	0.12
Backsweep Angle (deg)	68.7	Static Pressure Coefficient	0.28
TE to LE Radius Ratio	2.44	Stagnation Pressure Coefficient	0.32
TE to LE Span Ratio	0.27	Specific Speed	0.49
		Relative Velocity Diffusion Ratio	2.14

Table 6.2-1: Baseline Impeller Design Parameters

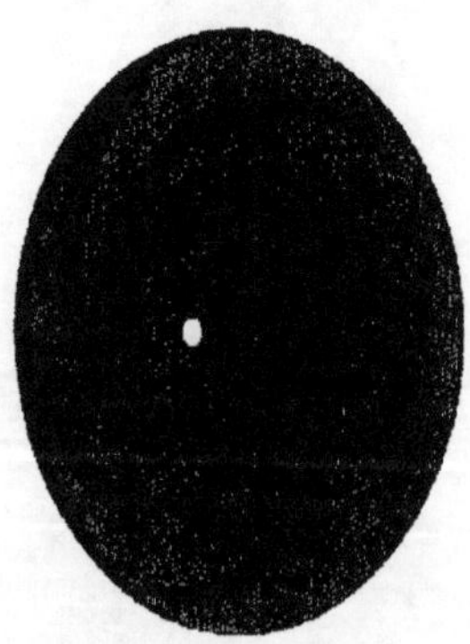

Fig. 6.2-1: View of the Baseline Impeller

Two features of the design are of importance for our study. First, the blade backsweep angle is high compared to the impellers in Hoshide, etal.[1971], Brasz [1988], and Ishida etal.[1981, 1990]. We will see in Section 6.6 that this aspect makes the impeller more sensitive to tip-clearance variations. The blade profile of the impeller is close to logarithmic[1] and the blade angle is almost constant from LE to TE.

Second, the impeller passage has an area decrease to roughly half the inlet area. Thus, as will be shown in detail later, the ratio of relative velocity at impeller exit to that at inlet is about 2.

We can place the impeller relative to parameters for current rocket pumps. Though most of the design parameters in Table 6.2-1 are in the current range for rocket applications [Pruger, 2001], we also need to consider the relative velocity diffusion ratio D defined as W_2/W_1. This ratio should be in the range 0.7-1.3, with current pump designs having a diffusion ratio close to 0.9 [Pruger, 2001]. It can thus be argued that the baseline pump has features not representative of a rocket pump.

The baseline impeller was therefore redesigned, with the meridional profile redrawn to change the area distribution and the radial profile modified by smoothing the trailing-edge profile to avoid excessive separation.

The final configuration thus had an outlet span of 2/3 of the inlet span, a relative velocity diffusion ratio of 1.3, but separation at outlet. Because of this separation, the analysis was confined to the first 80 % of the impeller passage from the leading edge plane, where there is no separation and the relative velocity diffusion ratio is about 0.9. The original and modified meridional profile are shown in Figure 6.2-3. Figure 6.2-4

---

[1] A purely logarithmic profile is shown in Figure 6.2-2. In this case, the blade angle β_{bm} from tangential is constant.

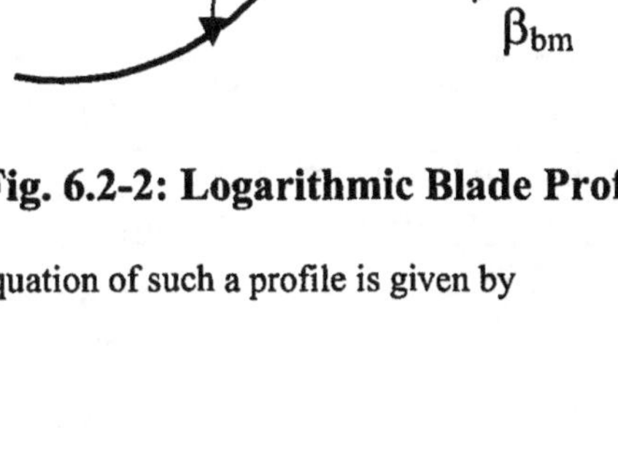

Fig. 6.2-2: Logarithmic Blade Profile

With r_1 the radius at inlet, the equation of such a profile is given by

$$\theta = \cot(\beta_{bm})\ln\frac{r}{r_1}. \tag{6.2-1}$$

shows a comparison between the baseline and modified casing of the pump, including the inlet pipe.

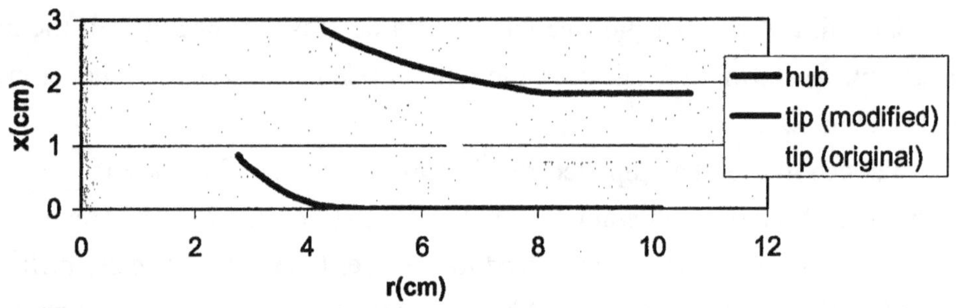

Fig. 6.2-3: Meridional Profiles of the Baseline and Modified Impellers

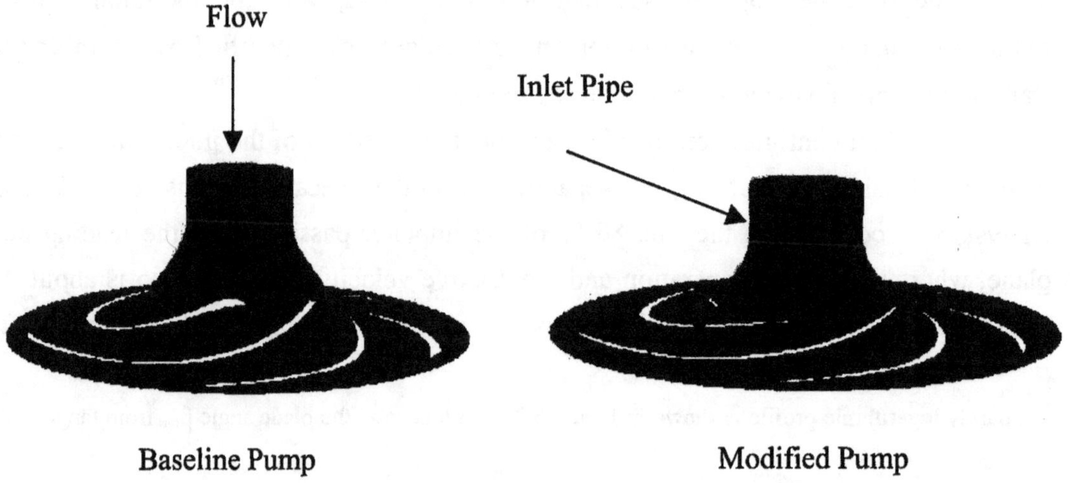

Fig. 6.2-4: Baseline and Modified Pump Casing

The design parameters for the modified pump (limited to first 80% of the impeller passage) are given in Table 6.2-2. This impeller is closer than the baseline one to the Rocketdyne impeller [Williams et al., 2001] in terms of specific speed and diffusion ratio.

Design Parameters	
Exit Flow Coeffcient	0.06
Static Pressure Rise Coefficient	0.41
Stagnation Pressure Rise Coefficient	0.66
Specific Speed	0.37
Diffusion Ratio	0.9

Table 6.2-2: Design Parameters of the Modified Pump

6.2.3 Implementation of Computational Procedure

6.2.3.1 Tip clearance Geometry

The tip clearance was created from the original shrouded design by reducing the blade height. The impeller casing remained the same for all cases. The physical tip clearance gap was kept constant along the impeller. This is representative of the practical situation in a centrifugal unshrouded impeller where tip clearance variations are induced by axial movements of the shaft [Prueger, 2001].

6.2.3.2 Computing the Flow Field

We used the GAMBIT meshing software [Fluent, 2001], which is provided in the FLUENT package, to generate the impeller unstructured grid. All the computations were run with the three-dimensional FLUENT solver (version 6) for incompressible flow with the use of a k-ε turbulence model.

6.2.3.3 Grid Generation

Since our investigation was focused on tip clearance effects, it was important to know how many nodes are required in the clearance region in order to have the tip clearance flow field appropriately resolved. Three levels of resolution were examined: three, five and nine nodes spanwise. The global results (pressure rise coefficients, flow averaged quantities), given in Table 6.2-3, were not sensitive to the mesh resolution; the sensitivity is below 1% when comparing the cases with 3,5 and 9 nodes for the modified impeller in the 5% tip clearance case. As shown in Section 6.5.2.1, the reduction in static and stagnation pressure rise coefficient due to a 5% increase in tip clearance in the modified impeller is around 10%. Hence, the differences in the estimations given in Table 6.2-3 are negligible.

Number of Nodes in the Tip Region	3	5	9	Maximum Difference (%)
Static Pressure Rise Coefficient	0.367	0.368	0.369	0.440
Stagnation Pressure Rise Coefficient	0.586	0.586	0.583	0.659

Table 6.2-3: Sensitivity of the Pressure Coefficient Evaluation to Tip Region Meshing (Modified Impeller, 5% tip clearance Case)

With regard to resolving the flow structure, the lowest resolution (three nodes) is not sufficient to capture the details in the tip-region. The two other resolutions appear to capture such details equally well (based on the examination of the reduced stagnation pressure contours). The flow in the clearance region has no high gradients in velocity or pressure that would require a tight meshing. There are high gradients near the wall but we are interested in the overall features of tip leakage flow (the inviscid leakage flow which flows over the blade tip and enters the channel), and do not intend to resolve the leakage jet boundary layer near the wall.

The main passage region away from the impeller tip region is less critical in terms of mesh requirements. For the blade surface, we chose a non-linear meshing, meaning that the interval between neighboring nodes is not constant, with more nodes in the vicinity of the tip than of the hub (interval ratio 1.1). Computations showed that the results are not sensitive to the blade spanwise resolution either. Table 6.2-4 shows that increasing the blade spanwise resolution by a factor of 2.5 only changes the computed pressure coefficient by a fraction of percent.

	Shrouded Case		Difference (%)	5% Tip Clearance Case		Difference (%)
Number of Nodes Spanwise on the Blade	12	30		12	30	
Static Pressure Rise Coefficient	0.410	0.409	0.26	0.368	0.366	0.54
Stagnation Pressure Rise Coefficient	0.663	0.660	0.45	0.586	0.584	0.34

Table 6.2-4: Sensitivity of the Pressure Coefficient Evaluation to Blade Spanwise Meshing Resolution (Modified Impeller)

When comparing results for different tip clearance ratios, the same meshing was used, to be sure the differences come from actual tip clearance effects and not from mesh resolution discrepancies. Figure 6.2-5 gives a sketch of the impeller meshing commonly used and Figure 6.2-6 shows details of the blade mesh.

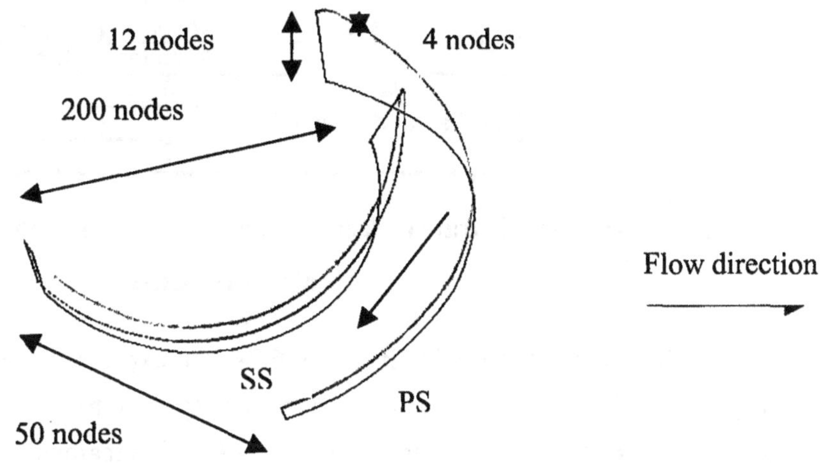

Fig. 6.2-5: Impeller Meshing

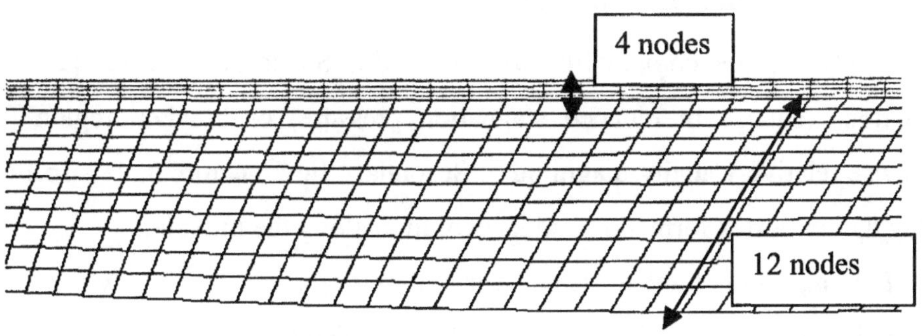

Fig. 6.2-6: Blade Meshing

6.2.3.4 Description of the Computations

According to Rocketdyne [Prueger, 2001], we could expect a 10% change in tip clearance during centrifugal pump operation, and the tip clearance simulations were chosen to remain in this range. The cases examined are presented in Table 6.2-5.

Cases	Baseline Impeller	Modified Impeller
	Shrouded	Shrouded
Tip Clearance Ratio (%)	0	0
	4	5
	8	

Table 6.2-5: Tip Clearance Cases Examined in Computations for Baseline and Modified Impeller

The tip clearance ratios in Table 6.2-5 are expressed with respect to the trailing-edge span. The 0% tip clearance case corresponds to a shrouded impeller with a stationary casing. The reason to investigate a 0% tip clearance is to isolate the effects of tip clearance from the effects of shroud/casing relative velocity (i.e. the difference between the shroud rotating with the blades or stationary) when we compare the different cases. This was found to be critical when comparing the secondary flow structure for different cases.

6.2.4 Interrogation of the Computed Flow Field

Analysis of the computations is presented in Section 6.3. Scalar quantities isolate loss regions (change in reduced stagnation pressure) or blocked region (change in streamwise relative velocity magnitude). Secondary flow vectors have also been plotted to study the tip-vortex formation and its evolution along the impeller.

It is also important to select a set of relevant surfaces on which to draw the contours lines and to visualize secondary flow. As detailed in Section 6.3.2.1, the choice of reference planes was orthogonal to the main flow direction in the shrouded case (see Figure 6.3-3).

6.3 Analysis and Synthesis of Computed Results

This Section describes the computed flow fields in a centrifugal pump impeller with tip clearances of 0, 4 and 8%. We first present the overall impact of tip clearance on performance in terms of performance sensitivity to tip clearance variation, at design point as well as at off-design. This is followed by linking the flow features associated with tip clearance to the observed performance sensitivity. Analyses of how the tip leakage flow develops in the impeller passage and of the tip vortex trajectory are also carried out.

6.3.1 Performance Sensitivity

6.3.1.1 Design Flow Coefficient

Table 6.3-1 gives the computed sensitivity for the baseline impeller at design flow coefficient.

	Stagnation Pressure Rise Sensitivity	Static Pressure Rise Sensitivity	Efficiency Sensitivity
0 to 4% Tip Clearance Ratio	2.96	2.27	0.65
0 to 8% Tip Clearance Ratio	2.98	2.65	0.68

Table 6.3-1: Baseline Impeller: Sensitivity at Design Flow Coefficient

Comparing the results with Table 6.1-1, the present impeller was more sensitive to tip clearance than impellers reported in Hoshide etal.[1972], Brasz [1988], Ishida etal.[1981, 1990], Goto [1992], Williams etal. [2001] (as far as stagnation pressure rise is concerned), but the efficiency sensitivity is in the range of the other pump data quoted. We will explain this result in Section 6.5.

The sensitivity for 0 to 4% tip clearance is similar to that between 4 to 8% tip clearance. The performance decrease can thus be considered as roughly linear, at least for the limited data, which is in accord with the trends reported in Ishida, etal.[1990] and Ishida etal. [1990].

6.3.1.2 Performance Curve

Figure 6.3-1 shows the results for static and stagnation pressure rise sensitivity at off design conditions. In Figure 6.3-1, the flow coefficient is defined as

$$\phi_2 = \frac{\overline{V_{r_2}}^A}{U_{tip}},$$

where V_{r2} is the radial velocity at the constant radius impeller exit surface and U_{tip} is the impeller rotational speed at exit. The operator $\overline{\ \ }^A$ means area average.

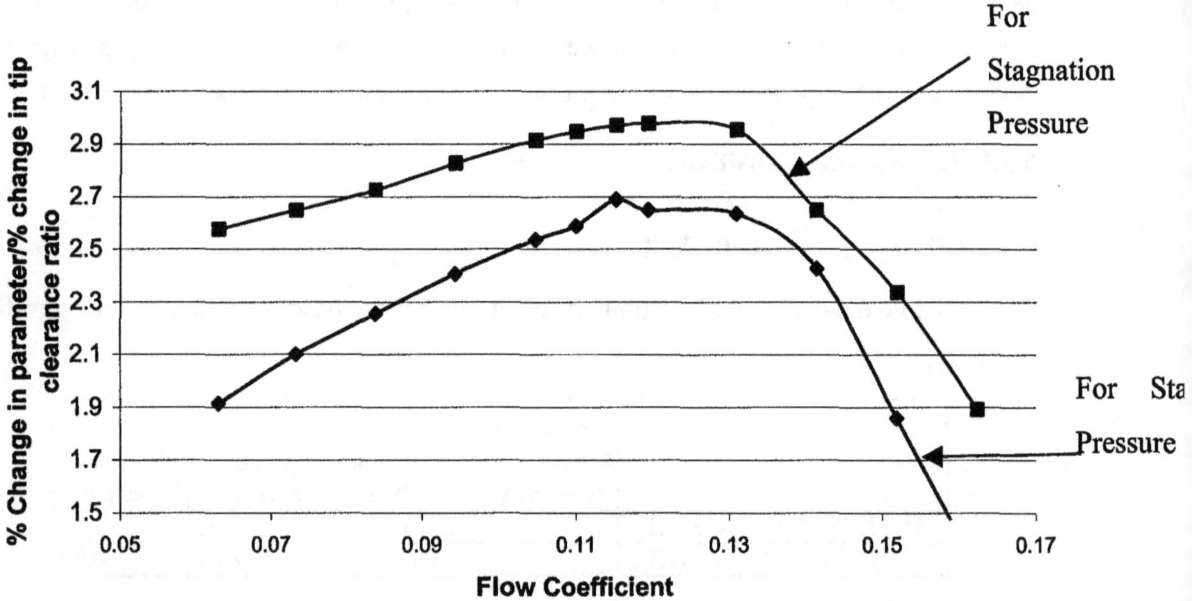

Fig. 6.3-1: Baseline Impeller: Variation of Pressure Rise Sensitivity with Flow Coefficient

Figure 6.3-1 shows the impeller is most sensitive at design conditions and the sensitivity decreases at operating points far from design conditions. The results for efficiency sensitivity, in Figure 6.3-2, show a similar trend. We will see in Section 6.6 that this trend can be captured using the simple flow model developed in Section 6.4.

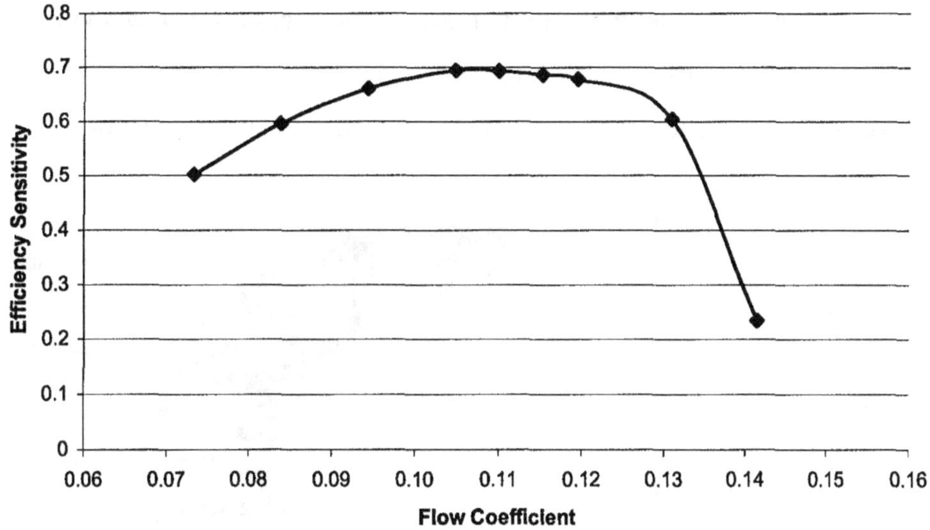

Fig. 6.3-2: Baseline Impeller: Variation of Efficiency Sensitivity with Flow Coefficient

6.3.2 Detailed Flow Field

6.3.2.1 Tip Clearance Flow Related Loss Generation and Blockage

The reduced stagnation pressure is the relevant quantity to examine loss generation in an impeller passage. The impeller inlet flow is uniform (it is delivered to the impeller by an axial duct) and the reduced stagnation pressure is conserved in the impeller in the absence of viscous effects. Reduced stagnation pressure contours such as those shown in Figures 6.3-4 and 6.3-5 therefore indicate regions in which entropy has been generated.

Blockage production can be seen from contours of relative velocity magnitude on suitably selected surfaces (principally planes orthogonal to the main flow direction) within the impeller passage. These identify the regions of low velocity. Since the wake is close to the suction side (see Figures 6.3-6 and 6.3-7), this velocity defect is not the result of local static pressure increase. Indeed, the suction side neighborhood is the region where the static pressure is minimum. Moreover, as will be seen in Section 6.3.2.2, tip leakage flow also causes a tip vortex structure at the casing/suction side corner. Due to the swirl, there is additional static pressure reduction in the region.

The same reference planes for contours are used for each tip clearance case. These planes, which are orthogonal to the flow in the shrouded case, are shown in Figure 6.3-3. These planes are nearly orthogonal to the main flow in the other tip clearance cases (their normal vectors differ by less than 3° from the tangential and axial direction).

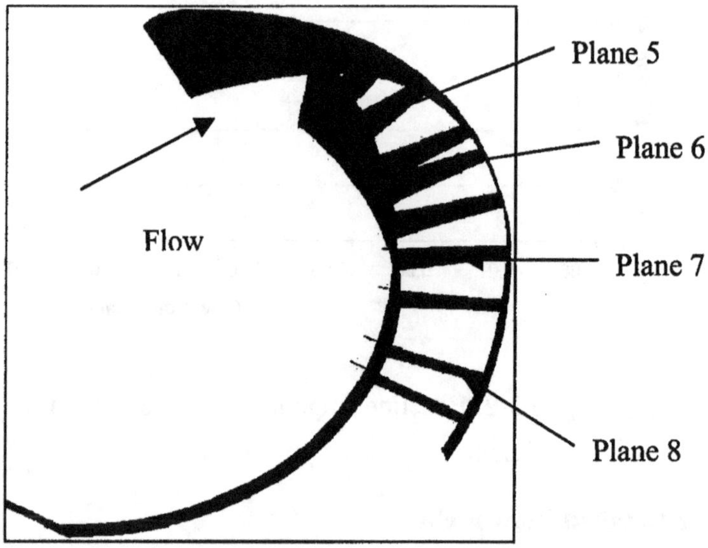

Fig. 6.3-3: Viewing Planes

6.3.2.1.1 Reduced Stagnation pressure

We define a non-dimensional reduced stagnation pressure by

$$CP_{Tred} = \frac{P_{T,red} - \overline{P_{T,red,inlet}}^{M}}{\rho U_{tip}^{2}}.$$

We focus on two points, namely the change in reduced stagnation pressure distribution with a change in tip clearance ratio and the streamwise evolution of the high-loss region.

Due to mixing of the tip leakage flow with the main flow, a high loss region develops at the casing/suction side corner. The extent of the region scales roughly linearly with tip clearance. The reduced stagnation pressure at the point where it is minimum also decreases with increasing tip clearance, as can be seen in Figure 6.3-4.

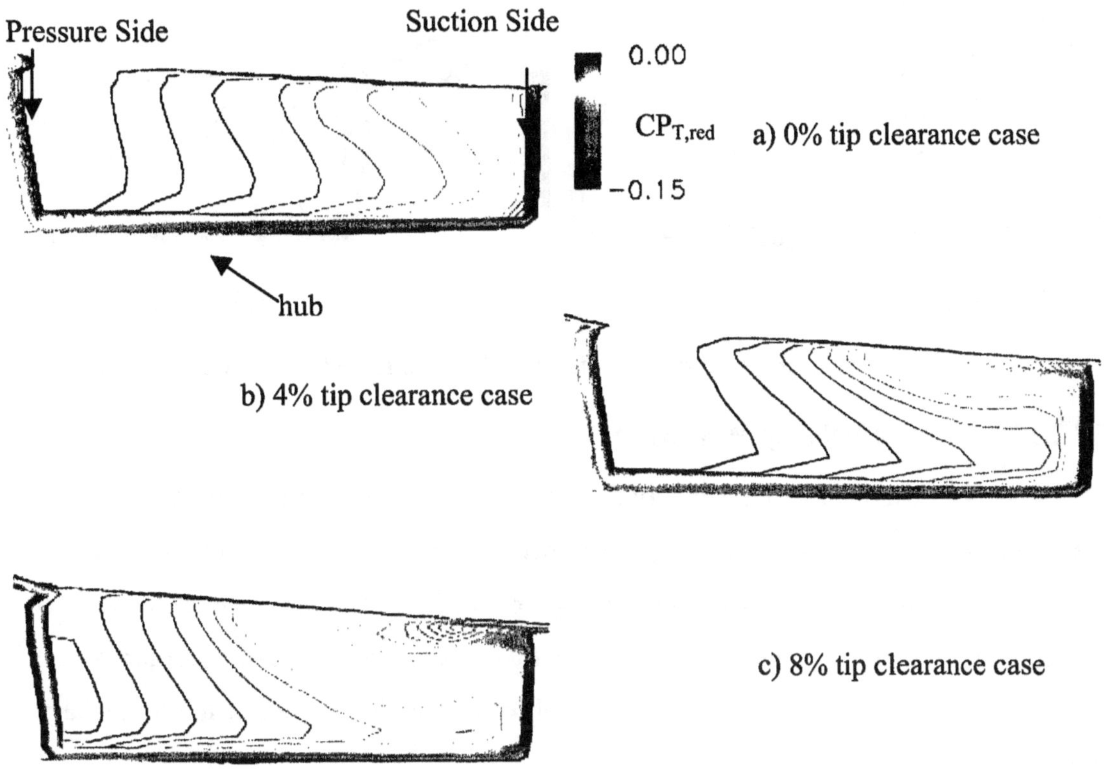

Fig. 6.3-4: Contours of $CP_{T,red}$ on Plane 8

The change in the distribution of reduced stagnation pressure as tip clearance is increased can be explained as follows. The mixing between tip leakage flow and main flow is not complete (this will be developed in Section 6.4), so the flow after mixing has lower reduced stagnation pressure than the main flow. As the clearance increases, there is more leakage flow and hence larger regions of low stagnation pressure.

Figure 6.3-5 shows that the high loss region expands downstream, growing pitchwise from the casing/suction side corner towards the center of the channel (see Figure 6.3-5). This corresponds to the accumulation of low energy tip leakage flow which has not completed mixing.

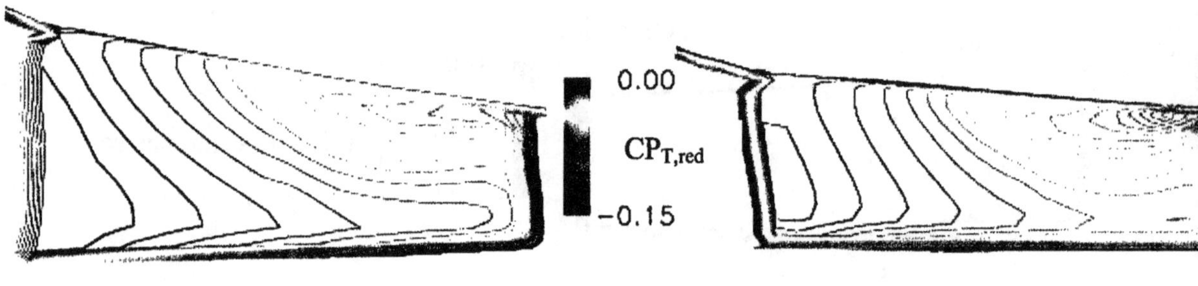

a) Plane 7 b) Plane 8

Fig. 6.3-5: Comparison of Contours of $CP_{T,red}$ in the 8% Tip clearance Case

6.3.2.1.2 Relative Velocity Magnitude

We can use the relative velocity distribution in the impeller passage to determine the blockage development due to tip clearance. Figure 6.3-6 shows relative velocity contours for an impeller with no tip clearance and with 8% tip clearance, at plane 7. A low relative velocity can be observed close to the suction side/casing corner of the impeller blade with clearance. This is the result of the partial mixing of the tip leakage flow with the main flow.

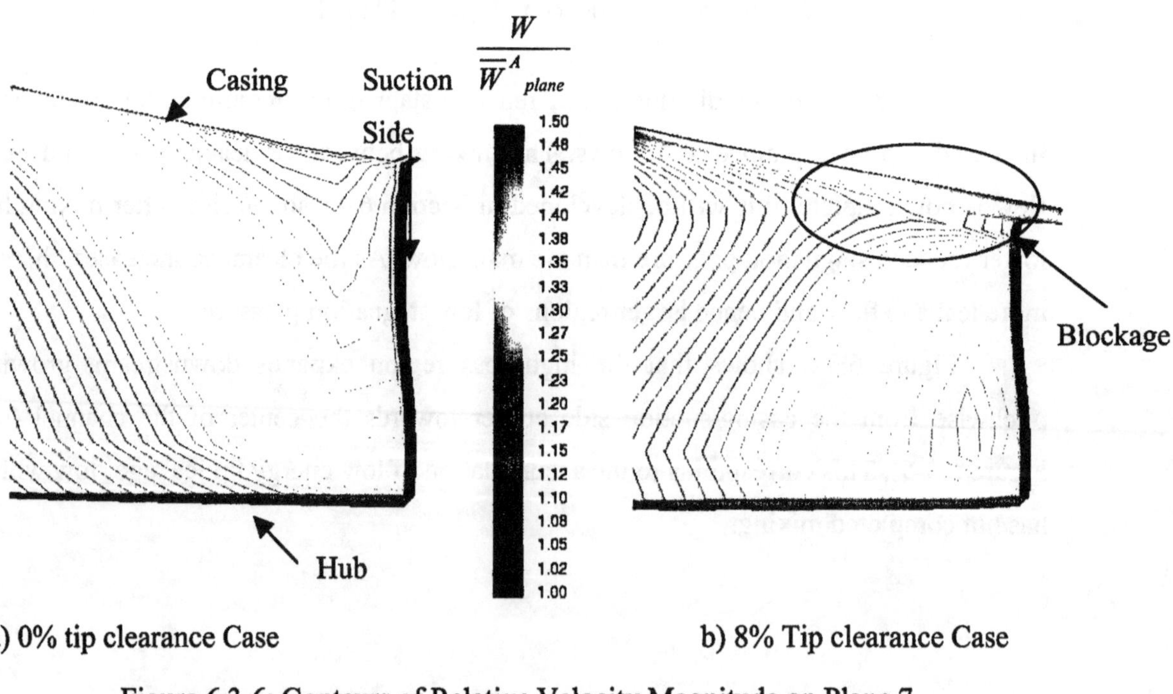

a) 0% tip clearance Case b) 8% Tip clearance Case

Figure 6.3-6: Contours of Relative Velocity Magnitude on Plane 7 (Normalized by the Average Velocity on Plane 7), Baseline Impeller

Comparison Figure 6.3-6 to Figure 6.3-7[1] shows the blocked region does not increase in extent or worsen downstream. This is expected because the main flow is accelerated; thus favorable reduced pressure gradient acts to reduce the relative velocity defect.

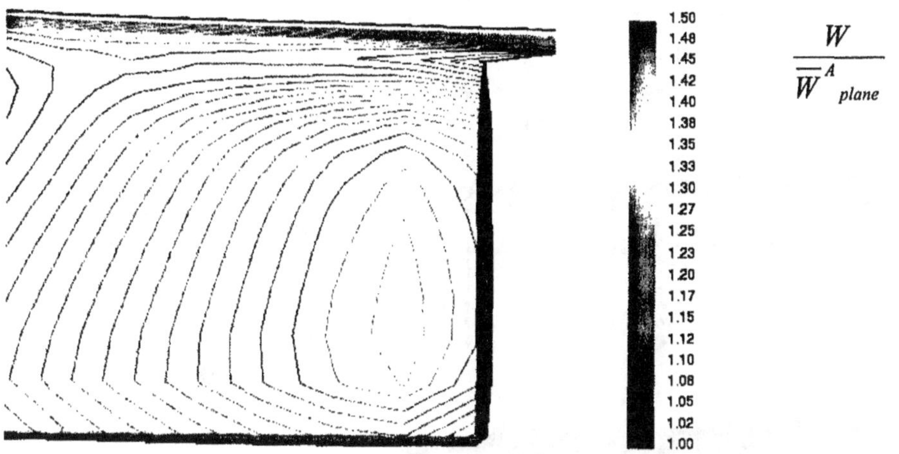

Fig. 6.3-7: Contours of Relative Velocity Magnitude on Plane 8 (Normalized by the Average Velocity on Plane 8), Baseline Impeller, 8% Tip Clearance Case

6.3.2.2 Secondary Flow Structure

It is also useful to examine the secondary flow structure and a method has been developed to visualize the secondary flows on arbitrary plane of reference . Figure 6.3-8 is a plot of secondary flow on Plane 8 for different tip clearance cases. The tip clearance vortex dominates the secondary flow structure for the 4% and 8% tip clearance cases. In the shrouded and 0% tip clearance case, the secondary flow is negligible except in the boundary layers. For the non-zero tip clearance cases, the shear layer that bounds the tip leakage flow rolls up into a vortical structure whose extent increases with clearance.

---

[1] The same scale is used for figure 6.3-6 and 6.3-7 but the value of $\overline{W}^{A}_{plane}$ is different for planes 7 and 8 due to the flow acceleration within the impeller. This explains the difference in the colors seen in the two figures.

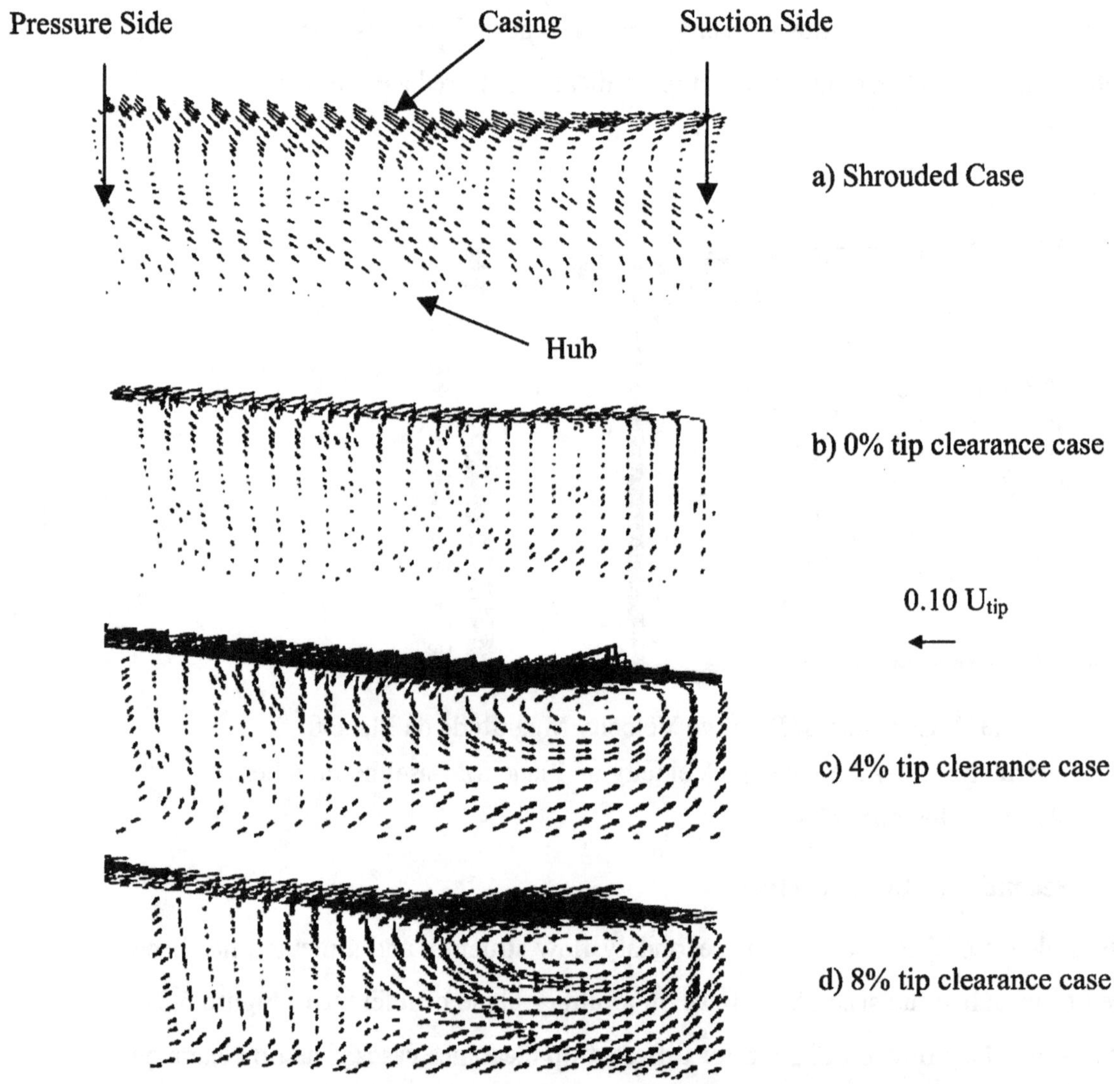

Fig. 6.3-8: Comparison of Secondary Flow Structure on Plane 8

Figure 6.3-9[2] shows the tip vortex size also increases with downstream distance. The vortex is hardly visible on plane 5 (close to the leading edge) but it grows to about a quarter of the impeller passage at exit by plane 8.

[2] There is no scale difference between Figures 6.3-8 and 6.3-9; only the size of the figures is different.

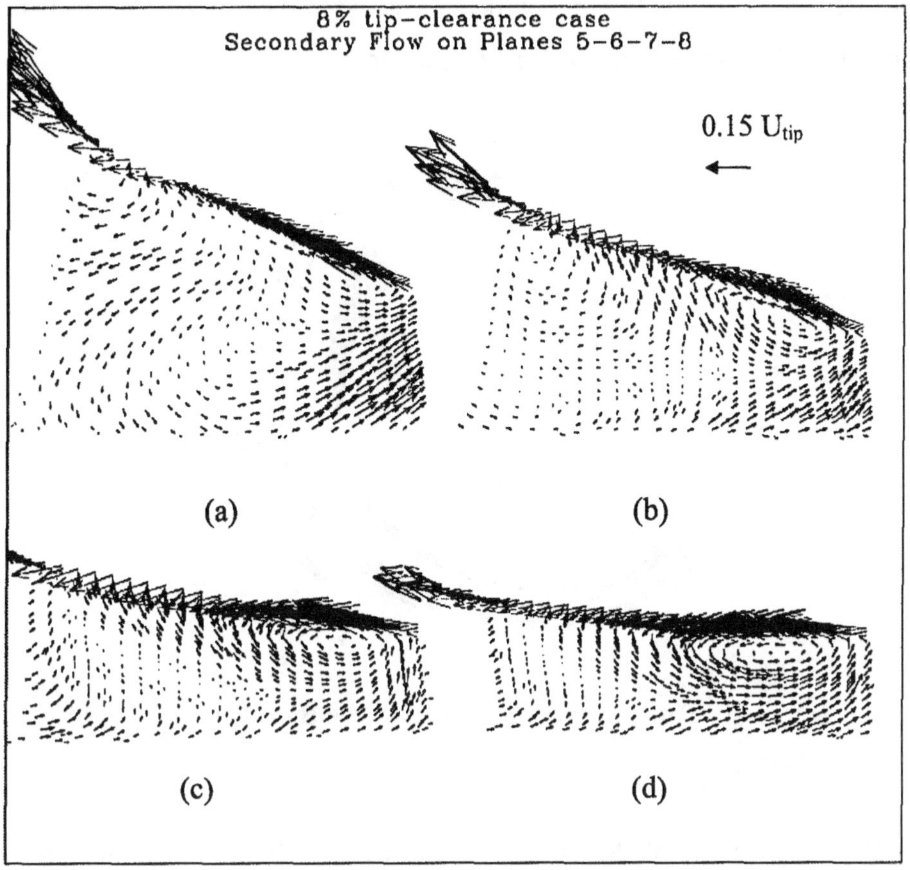

Fig. 6.3-9: Streamwise Evolution of the Secondary Flow Structure in the 8% Tip Clearance Case: (a) Plane 5, (b) Plane 6, (c) Plane 7, (d) Plane 8

6.3.2.3 Correspondence between the Low Stagnation Pressure region and the Tip Vortex Region

Comparing Figures 6.3-4 and 6.3-5 to Figures 6.3-9 and 6.3-10 indicates that reduced stagnation pressure contours show a similar evolution as the vortex vector plots. We now intend to show that the low energy tip leakage flow is embedded in the tip vortex region, and that the region of low reduced stagnation pressure effectively coincides with the tip vortex. A consequence is that the tip leakage flow does not achieve complete mixing with the main flow. Moreover, the mean trajectory of the tip leakage streamlines can be identified with the trajectory of the tip vortex center (see Section 6.3.2.4).

Tracking the tip leakage flow in the channel enables us to support the statement that tip leakage flow is embedded in the tip vortex. Figure 6.3-10 shows a set of tip leakage streamlines for the baseline impeller with 8% tip clearance. We can see that the leakage lines roll up and spiral in a vortex structure.

Fig. 6.3-10: Tip Leakage Streamlines in the Baseline Impeller, 8% Tip Clearance Case.

We now track sample leakage streamlines and study their position with respect to the vortex core and the low reduced stagnation pressure region. Figure 6.3-11 shows how the tip leakage streamlines intersect the tip vortex on planes 7 and 8, for the baseline impeller with 8% tip clearance. Figure 6.3-12 shows how the tip leakage streamlines intersect the low reduced stagnation pressure region on the two same planes. While the streamlines do not intersect at the point of lowest reduced stagnation pressure (close to the casing/suction side corner, see line contours on Figure 6.3-5), they pass through the high loss region which coincides with the vortex core. We have chosen to plot filled contours for visibility reasons.

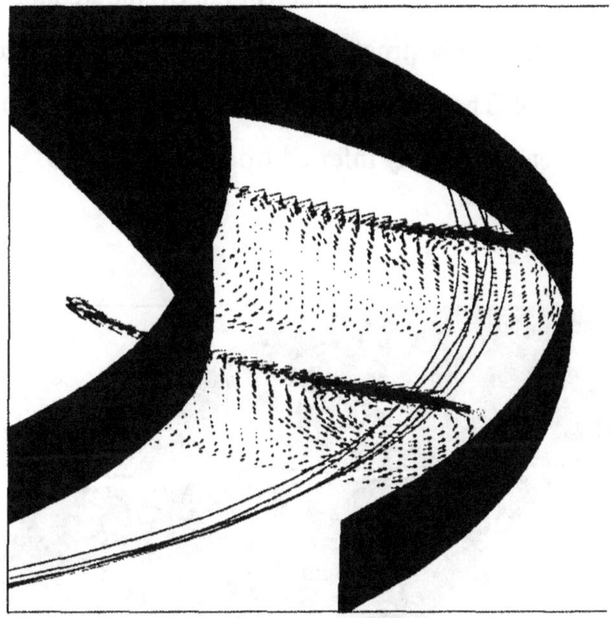

Fig. 6.3-11: Tip leakage Streamlines Embedded in the Tip Vortex in the Baseline Impeller, 8% Tip Clearance Case

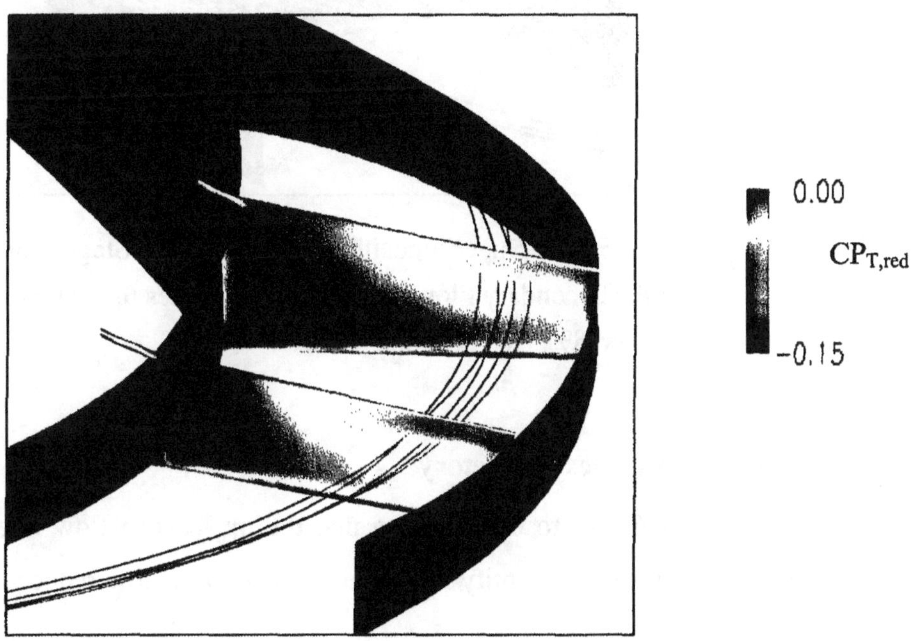

Fig. 6.3-12: Tip leakage Streamlines Intersecting the Loss Region in the Baseline Impeller, 8% Tip Clearance Case

Figure 6.3-13 shows the superimposition of the reduced stagnation pressure contours and the secondary flow structure on planes 6,7 and 8, in the baseline impeller, in the 8% tip clearance case. The overlapping of the low reduced stagnation pressure region with tip vortex region can be readily inferred from Figure 6.3-13.

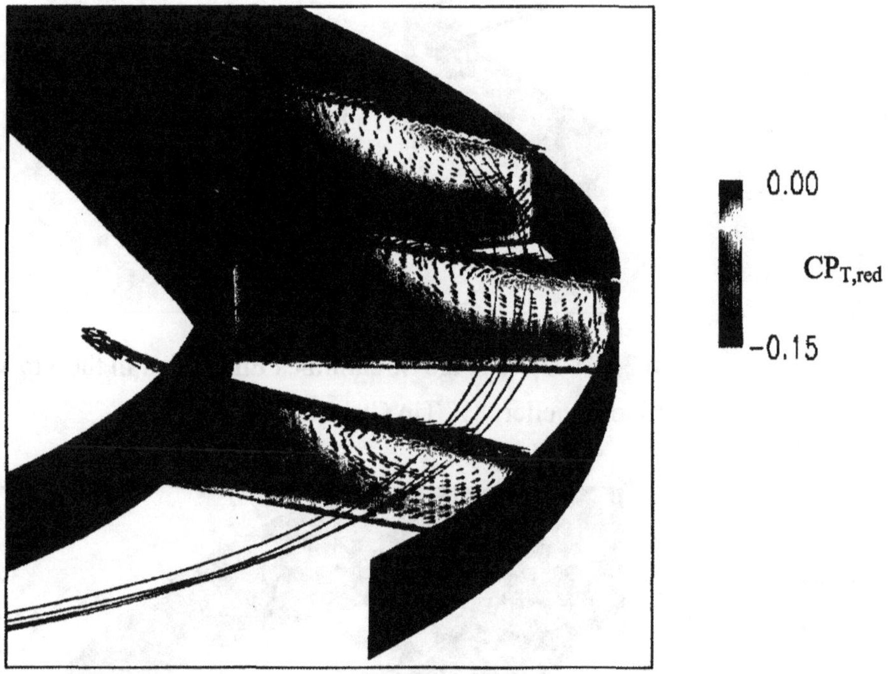

Figure 6.3-13: Superimposition of the Reduced Stagnation Pressure Contours and the Secondary Flow Vector Plot on Planes 6, 7, and 8 in the Baseline Impeller, 8% Tip Clearance Case

6.3.2.4 Tip Vortex Trajectory

Figures 6.3-10 to 6.3-13 show that the tip leakage flow is embedded in the tip vortex and that we can identify the vortex core with the low reduced stagnation pressure region.

Analyzing the tip vortex trajectory can therefore help in presenting a picture of blockage evolution. To accomplish this, the approach of Chen, etal. [1991] has been adapted for a centrifugal impeller.

For axial compressors, the trajectory of the tip clearance vortex center can be shown to be a function of a single non-dimensional parameter t* defined as

$$t^* = \frac{t}{\tau}\sqrt{\frac{\Delta P}{\rho}},$$

where ΔP is the pressure difference across the blade, τ the tip clearance gap, ρ the density, and t the convective time (streamwise distance divided by velocity).

The analysis can be extended for a centrifugal machine and the same definition of t^* is used. The corresponding expression for t* in terms of impeller geometry and flow parameters is given by

$$t^* = \frac{(r - r_1)}{\tau}\sqrt{\frac{2\pi}{Z}(1-\chi)^{-1}\left[\frac{1-\chi^2}{\phi_2} - \tan\beta_2 + \chi\tan\beta_1\right]}, \quad (6.3\text{-}1)$$

where Z is the number of blades and $\chi = \frac{r_1}{r_2}$ the inlet to outlet radius ratio .

The non-dimensional vortex center coordinates, which are only a function of t*, are:

- $y_c^* = y_c/\tau$, with y_c the distance between the vortex center and the blade
- $z_c^* = z_c/\tau$, with z_c the distance between the vortex center and the casing

Figure 6.3-14 shows the results for y_c^* for the 4% and 8% tip clearance cases. The points on the figure represent the coordinates (t^*,y_c^*) of the vortex center measured from the CFD results, and the line is a straight line fit to the data.

The equation of the curve is

$$y_c^* = 0.0754t^*. \quad (6.3\text{-}2)$$

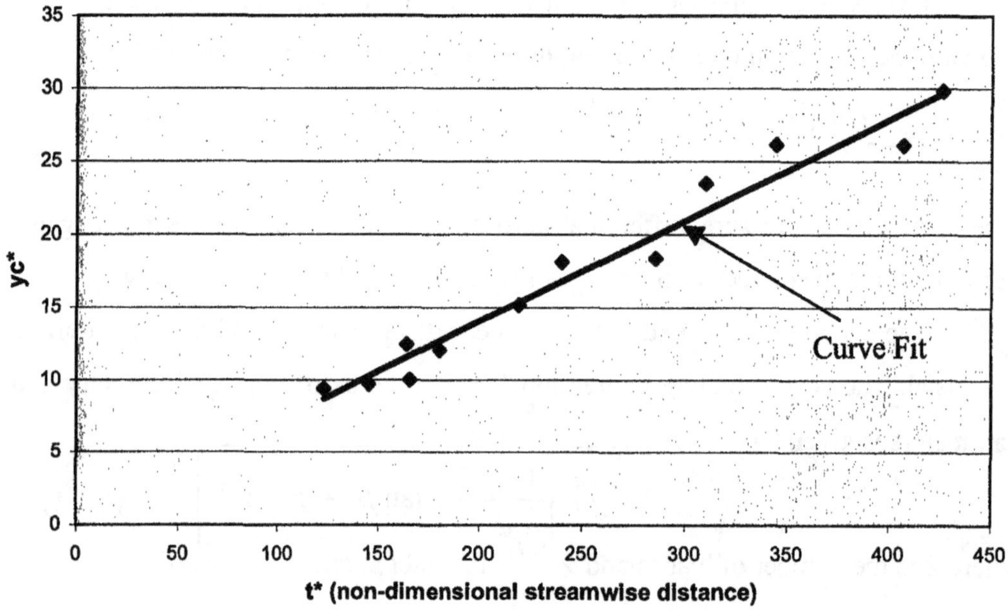

Fig. 6.3-14: Tip Vortex Center Trajectory (Pitchwise Coordinate)

As with axial compressors, there is a linear relation between y_c* and t*. From this analysis, one can predict the trajectory of the tip vortex center knowing the pump design parameters and thus estimate the wake trajectory, which is an important element in evaluating the effects of blockage on performance (see Section 6.4.3.2).

6.3.3 Summary

Computations show that the sensitivity of pressure rise to tip clearance is a maximum at design conditions. The percent change in efficiency is lower than for pressure rise but is also maximum at design conditions. The computations show that the tip leakage flow partially mixes with the main flow when it enters the channel, creating a region of low reduced stagnation pressure. The resulting low velocity flow then proceeds downstream, embedded in the tip vortex structure. An analysis of the tip vortex trajectory shows that non-dimensional vortex center coordinates are a linear function of a single non-dimensional parameter t*.

6.4 Flow Modeling for Tip Leakage Flow

This Section details the development of a model for the effects of tip leakage on centrifugal impeller performance. First, an analysis of the phenomena that are known to change the performance is implemented. In this we distinguish between mixing, blockage, and their impact on pressure rise. We then detail the procedure for choosing a mixing model and present expressions for mixing losses in the impeller. A blockage estimate, developed for axial compressors [Khalid et al., 1999] is also adapted for use in centrifugal impellers to provide estimates of performance sensitivity to tip clearance as function of different design parameters.

6.4.1 Introduction: Physical Phenomena Involved

6.4.1.1 Description

Two physical processes are associated with the observed performance decrease due to tip clearance. First, the tip leakage flow entering the channel mixes with the main flow. This constitutes a loss. Second, once mixing has occurred, the leakage flow consists of a region of low streamwise velocity compared to the core flow, and hence there is blockage generation. The blockage leads to a drop in both stagnation and static pressure rise capability in the impeller. Static pressure is affected by a reduction of effective through-flow area. Stagnation pressure drops due to a change in exit absolute tangential velocity. This change corresponds to a deficit of work input and is not associated with loss generation although the subsequent mixing is.

In the next Section an expression is presented linking the change in static and stagnation pressure rise due to blockage and mixing loss associated with impeller tip leakage flow.

6.4.1.2 Contributions for Stagnation and Static Pressure Rise

6.4.1.2.1 Stagnation Pressure Rise

For an adiabatic process across the impeller,

$$\frac{\Delta \overline{P_t}^{M}}{\rho} = W_{shaft} - T_t \Delta \overline{s}^{M} . \qquad (6.4\text{-}1)$$

From the Euler Turbine Equation, we have, neglecting the viscous forces from shear on the casing,

$$W_{shaft} = \Delta(U\overline{V_\theta}^M). \tag{6.4-2}$$

We can estimate the magnitude of viscous forces by comparing the stagnation pressure coefficient for the shrouded and 0% tip clearance case; these differ by less than 1%. The stage pressure difference due to clearance is 3% for each percent of tip-clearance ratio increase. It is thus a good approximation to neglect the viscous forces from shear on casing.

Combining Equations 6.4-1 and 6.4-2, and non-dimensionalizing, we find

$$CP_T = \frac{\Delta\left(U\overline{V_\theta}^M\right)}{U_{tip}^2} - P_{loss}, \tag{6.4-3}$$

where $P_{loss} = \frac{T_t \Delta \overline{s}^M}{U_{tip}^2}$ is the non-dimensional entropy generation coefficient. There are two sources of stagnation pressure rise decrease, the reduction in work input which is related to blockage (change in absolute exit tangential velocity) and the entropy generation due to irreversibility which constitutes a loss.

6.4.1.2.2 Static Pressure Rise

Both mixing and blockage influence the static pressure rise. Equation 6.4-1 can be written as

$$\frac{\Delta \overline{P_S}^M}{\rho} = \frac{\Delta U^2}{2} - \frac{\Delta \overline{W^2}^M}{2} - T_t \Delta \overline{s}^M. \tag{6.4-4}$$

In non-dimensional form, with $CP_{centrifugal} = \frac{\Delta U^2}{2\rho U_{tip}^2}$, this is

$$CP_S = CP_{centrifugal} - \frac{\Delta \overline{W^2}^M}{2U_{tip}^2} - P_{loss}. \tag{6.4-5}$$

Mixing reduces static pressure rise through entropy generation. Blockage reduces static pressure rise through a reduction in effective area and hence a reduction in the diffusion term ΔW^2. The first right-hand term in Equation 6.4-5 $\frac{\Delta U^2}{2}$ corresponds to the pressure rise linked to centrifugal effect which is a loss free term.

6.4.2 Choice of a Mixing Model

In Section 6.1, two different models for mixing were described, the loss of cross flow energy model and the control volume analysis of Storer and Cumpsty. From the computational results, the former is more appropriate here because the mixing of the tip leakage flow with the core flow is only partial. As seen in Section 6.3, a low reduced stagnation pressure region develops at the suction side/casing corner where the tip leakage flow enters the channel. This region grows downstream, indicating that the tip leakage flow does not complete mixing.

Section 6.1 provided expressions for the local entropy generation. We thus need to develop an expression that applies to the whole domain. An estimate for the tip leakage velocity is given below as Equation 6.4-6. This result which has been used in Senoo et al.[1985, 1991] and Denton [1993] is also supported by the CFD results.

$$W_n = \sqrt{\frac{2(P_P - P_S)}{\rho}} = U_{tip}\sqrt{2CP_{s,loading}}\,. \tag{6.4-6}$$

The coefficient CP_S is defined as

$$CP_{S,loading} = \frac{P_p - P_s}{\rho U_{tip}^{\,2}}\,. \tag{6.4-7}$$

Assuming the blade thickness is small, we can write

$$W_n = \sqrt{W_S^{\,2} - W_P^{\,2}}\,, \tag{6.4-8}$$

where W_S and W_P are evaluated at the same radius.

The flow through the tip clearance is a vena contracta. As such we have to include a contraction coefficient C_d. The value can be derived from CFD results, or experimentally, as in the range 0.7-0.8. We used C_d=0.71. The leakage flow rate across blade length *dl* is given by

$$d\dot{m}_c = C_d \rho \tau W_n dl\,. \tag{6.4-9}$$

Hence

$$\dot{m}_m T\Delta s = \frac{1}{2}\int_{s_1}^{s_2} C_d \rho \tau W_n^{\,3} dl \,. \qquad (6.4\text{-}10)$$

We now want to make this expression non-dimensional. The mainstream mass flow is

$$\dot{m}_m = \frac{2\pi r_2}{Z} h_2 V_{r_2} \,, \qquad (6.4\text{-}11)$$

where h_2 is the outlet blade span, and V_{r2} is the radial velocity at outlet. We introduce the flow coefficient, $\phi_2 = \frac{V_{r_2}}{U_{tip}}$, the tip clearance ratio $\lambda = \frac{\tau}{h}$ and make use of the logarithmic profile geometry,

$$dl = \frac{dr}{\sin(\beta_b)} \,. \qquad (6.4\text{-}12)$$

Using the definition of P_{loss} in Section 6.4.1.2.1, Equation (6.4-10) becomes

$$P_{loss} = \frac{Z}{2\pi r_2 \phi_2} \int_{r_1}^{r_2} \frac{C_d \lambda}{\sin(\beta_b)} \left(\frac{W_n}{U_{tip}} \right)^2 \frac{W_n}{2U_{tip}} dr \,. \qquad (6.4\text{-}13)$$

6.4.3 Blockage Model

6.4.3.1 Evolution of a Velocity Defect through a Pressure Gradient

We now address the clearance-related blockage generation. To evaluate the reduction in work input due to blockage, we adapt a simplified version of the method used by Khalid et al. [1991] for the evolution of a velocity defect in a blade passage.

We wish to examine the evolution of the blocked region area in a (reduced) static pressure gradient. We consider a quasi-one-dimensional, inviscid, incompressible flow with no mixing. This assumption is based on the CFD results which show the velocity defect region is embedded in a roll-up vortex structure: this limits the mixing with the surrounding flow. The quasi-one-dimensional approximation is possible because the flow has an easily identifiable main-direction in the channel.

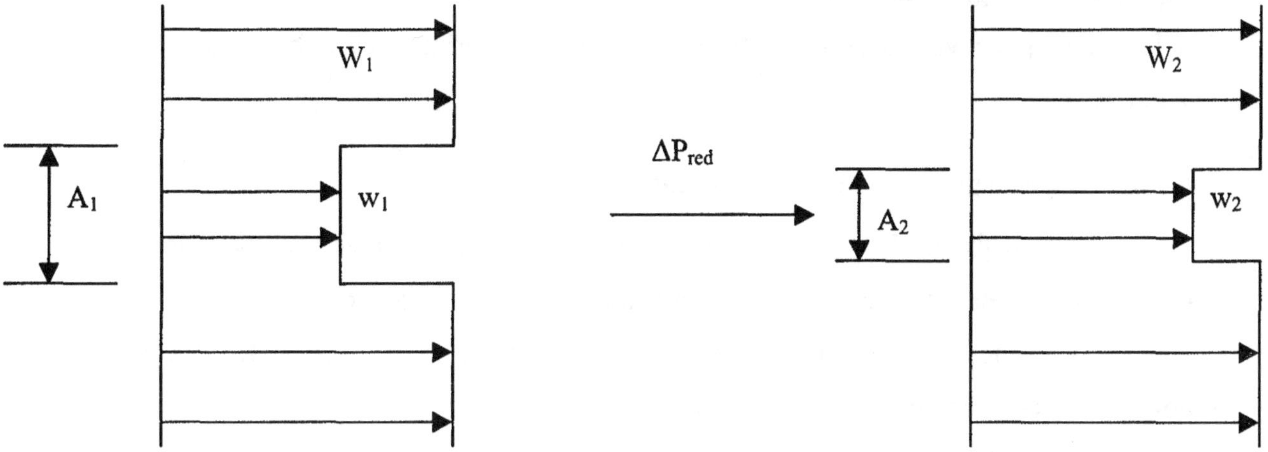

Fig. 6.4-1: Evolution of a Velocity Defect in a Reduced Pressure Gradient

Figure 6.4-1 describes the situation. At station 1 (left part), there is a region of low velocity that occupies an area A_1. The blockage at station 2 is

$$A_{b2} = \left(1 - \frac{w_2}{W_2}\right) A_2 . \tag{6.4-14}$$

From continuity in the defect region,

$$A_1 w_1 = A_2 w_2 . \tag{6.4-15}$$

Let γ_1 denote $\frac{w_1}{W_1}$ the initial defect ratio. With no mixing, conservation of reduced stagnation pressure in the defect and the core flow yields

$$\frac{w_2}{W_1} = \sqrt{\gamma_1^{\ 2} - CP_{red}} \ , \ \frac{W_2}{W_1} = \sqrt{1 - CP_{red}} \ \text{ where } CP_{red} = \frac{\Delta P_{red}}{\frac{1}{2}\rho W_1^{\ 2}} . \tag{6.4-16}$$

The blocked area is

$$\frac{A_{b2}}{A_1} = \left(\frac{W_1}{w_2} - \frac{W_1}{W_2}\right) \frac{w_2}{W_1} \frac{A_2}{A_1} . \tag{6.4-17}$$

Combining Equations 6.4-14 to 6.4-17:

$$\frac{A_{b2}}{A_1} = \gamma_1 \left[\left(\gamma_1^{\ 2} - CP_{red}\right)^{-1/2} - \left(1 - CP_{red}\right)^{-1/2} \right] . \tag{6.4-18}$$

Equation 6.4-18 gives the blocked area at station 2 as a function of the initial velocity defect ratio γ_1, the wake area A_1 and the reduced pressure variation CP_{red} between station 1 and 2.

6.4.3.2 Leakage Flow Blockage Estimation

To find the total blockage, it is assumed that we can "break up" the velocity defect into discrete wakes corresponding to neighboring chordwise locations. Each of the local wake generated by tip leakage flow is supposed to proceed independently downstream (see Figure 6.4-2).

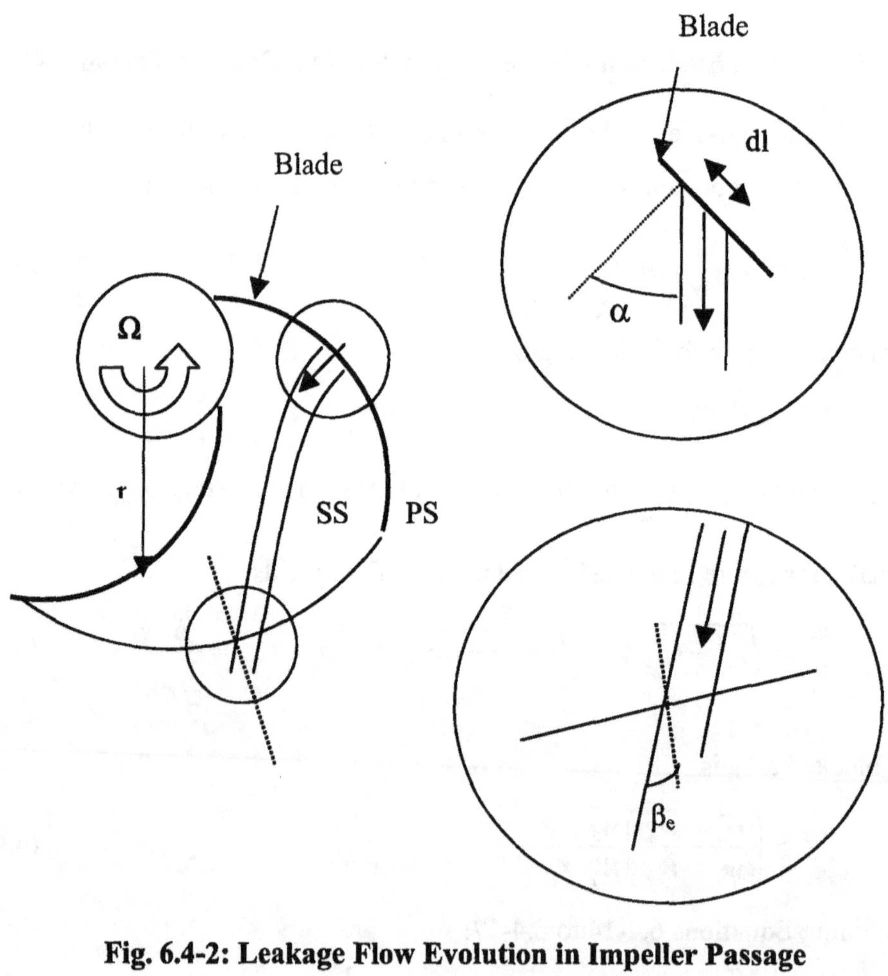

Fig. 6.4-2: Leakage Flow Evolution in Impeller Passage

The description of the leakage flow evolution in the impeller passage consists of three elements:

1. For each chordwise element, the local tip leakage flow rate is derived from the local blade loading (see Equations 6.4-6 to 6.4-9). The initial wake area after loss of cross-flow energy is derived from the tip leakage angle relative to the main-flow

   $$A_{wake} = dl.\tau.C_d.\cos\alpha \,. \qquad (6.4\text{-}19)$$

   In Equation 6.4-19 $\alpha = \sin^{-1}\left(\frac{W_P}{W_S}\right)$, τ is the clearance gap, and C_d is the contraction factor

   The initial defect ratio, which appears in the defect evolution model is

   $$\gamma_1 = \frac{W_P}{W_S} \,. \qquad (6.4\text{-}20)$$

   where W_p is the pressure side velocity magnitude and W_S is the suction side velocity magnitude.

2. Each discrete wake proceeds downstream in the reduced pressure gradient. The effective blocked area $A_{b,exit}$ at impeller exit for the element is found using Equation 6.4-18.

3. The wake direction β_e is found from the tip vortex center trajectory analysis of Section 6.3 and yields the effective blocked area projected on constant radius surface at exit.

   $$A_{b,r2} = \frac{A_{b,exit}}{\cos(\beta_e)} \,. \qquad (6.4\text{-}21)$$

We then add the contributions of the individual discrete wakes to get the total effective blocked area at impeller exit.

6.4.3.3 Effect of Blockage on Stagnation Pressure Rise

We now evaluate the effect of blockage on stagnation pressure rise. We are interested in stagnation pressure change associated with the change in work input, and we thus begin with the Euler Turbine Equation

$$H = 1 - \phi_2 \tan\beta_2 \qquad (6.4\text{-}22)$$

with H the work input coefficient, $H_{Euler} = \frac{W_{shaft}}{\rho U_{tip}^{\;2}}$. Equation 6.4-22 relates the work input to the exit absolute tangential velocity for the case of no inlet absolute tangential velocity.

At exit, on a mass average basis

$$H_{Euler} = 1 - \overline{\phi_2 \tan\beta_2}^M . \qquad (6.4\text{-}23)$$

Based on the CFD results, we can write the mass average of the product in Equation 6.4-23 as the product the mass averages of the two terms with a 2% error (based on the mass average of the product):

$$\overline{\phi_2 \tan\beta_2}^M \approx \overline{\phi_2}^M \, \overline{\tan\beta_2}^M . \qquad (6.4\text{-}24)$$

Table 6.4-1 shows the comparison of the two estimates for the baseline and the modified impeller. The two terms of Equation 6.4-24 are close due to the fact that the flow angle is uniform at impeller exit. The simplification in Equation 6.4-24 enables us to obtain analytical expressions and will be used in the remaining of the thesis.

	$\overline{\phi_2 \tan\beta_2}^M$	$\overline{\phi_2}^M \, \overline{\tan\beta_2}^M$	Difference (%)
Baseline Impeller	0.611	0.625	2.24
Modified Impeller	0.393	0.399	1.50

Table 6.4-1: Averaging Approximation

The blockage affects the flow coefficient through reduction of the effective area. The blockage effect on stagnation pressure rise is due to the work reduction. Hence, the stagnation pressure rise change due to blockage is

$$\Delta CP_{T,blockage} = \overline{\tan\beta_2}^M \, \Delta\overline{\phi_2}^M . \qquad (6.4\text{-}25)$$

But:

$$\Delta\overline{\phi_2}^M = \frac{\Delta\overline{V_{r,2}}^M}{U_{tip}} = \frac{-\overline{\cos(\beta_2)}^M \, \Delta\overline{W_2}^M}{U_{tip}} . \qquad (6.4\text{-}26)$$

Combining Equations 6.4-25 and 6.4-26:

$$\Delta CP_{T,blockage} = -\overline{\tan\beta_2}^M \frac{\Delta\overline{W_2}^M}{\overline{W_2}^M} \overline{\phi_2}^M . \qquad (6.4\text{-}27)$$

We can define three contributions to stagnation pressure coefficient variation due to blockage:

1. The exit relative flow angle β_2. The larger the angle, the higher the effect of blockage on stagnation pressure rise.
2. The velocity variation which is a function of the blocked fraction of the exit area

$$\frac{\overline{\Delta W_2}^M}{\overline{W_2}^M} = \frac{A_{b,r2}}{A_{r2} - A_{b,r2}} \text{ (from continuity).} \qquad (6.4\text{-}28)$$

The blocked area is a function of blade loading, tip clearance ratio and streamwise reduced pressure gradient.

3. The mass average flow coefficient at exit.

6.4.3.4 Effect of Blockage on Static Pressure Rise

From Equation 6.4-5, the blockage is related to the static pressure rise by

$$CP_{S,diffusion} = -\frac{\Delta W^2}{2U_{tip}^2}. \qquad (6.4\text{-}29)$$

For small changes in W_2,

$$\Delta CP_{S,blockage} = -\left(\overline{\cos\beta_2}^M\right)^{-2} \frac{\overline{\Delta W_2}^M}{W_2^M} \left(\overline{\phi_2}^M\right)^2. \qquad (6.4\text{-}30)$$

The three contributions in the expression for the stagnation pressure rise change due to blockage also appear in the expression for the static pressure rise change, but with different importance:

1. The exit relative flow angle β_2. The higher the angle, the higher the effect of blockage on static pressure rise.
2. The velocity variation.
3. The flow coefficient at exit.

We will examine the parametric dependence of these in Section 6.6.

6.4.4 Model Summary

A summary of how the model is implemented is given in Figure 6.4-3. The required inputs are showed in the top box. The left wing shows the estimation of blockage, the right shows the estimation of mixing. For each contribution, the successive steps of the analysis are given as well as which tools (blue text) are used.

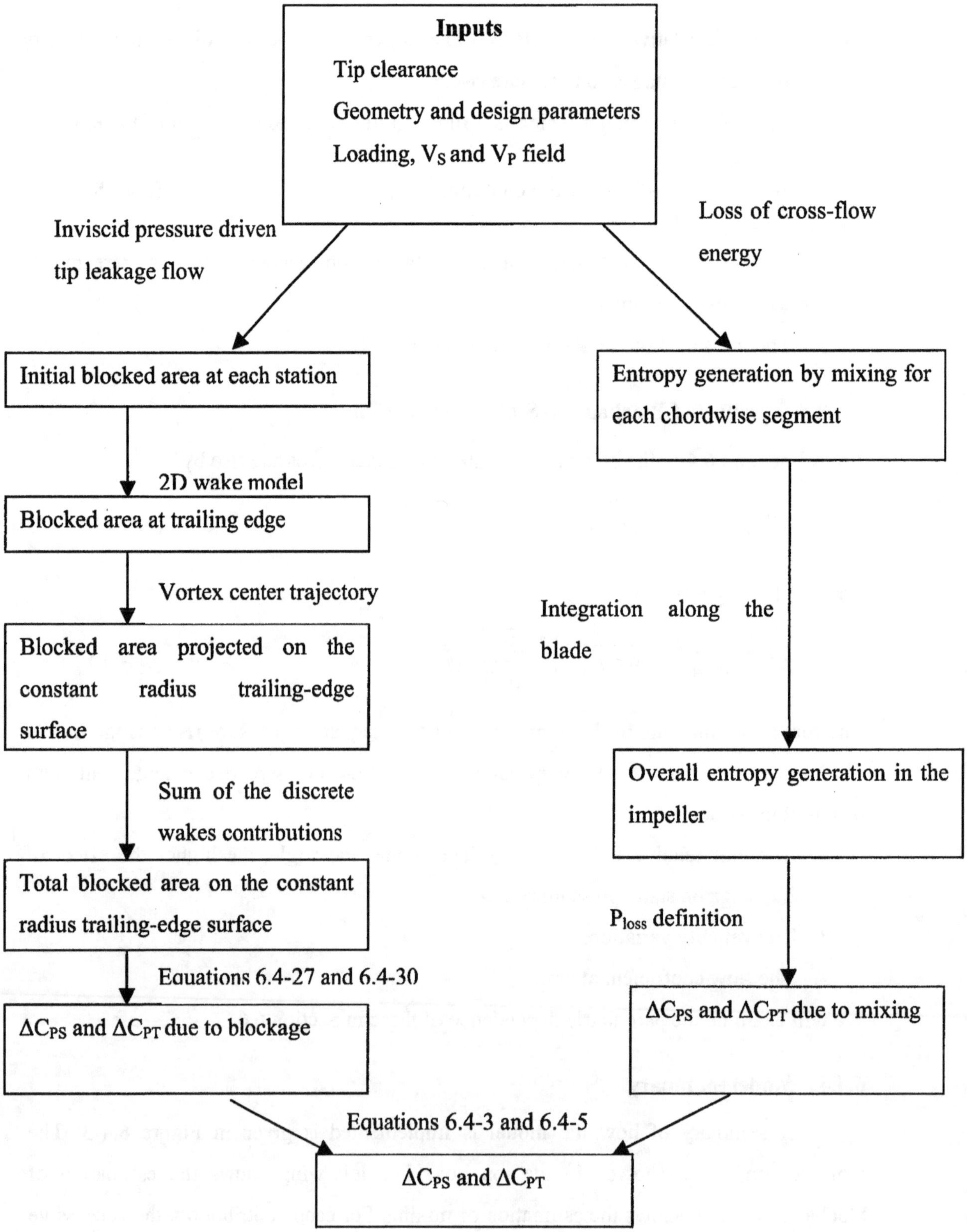

Figure 6.4-3: Model Implementation

6.5 Assessment and Applications of the Flow Model

One issue still to be addressed is how the model in Section 6.4 should be interpreted for a stage rather than an isolated impeller. This is examined below. The model[1] is then assessed using the computed results for the baseline and modified pump impeller. Performance parameters considered are static and stagnation pressure rise coefficient, efficiency, and blade loading variation (measured as the pressure difference across the blade). To apply the model to published results, we also define the information required for its use and assess the model against the information from representative impellers reported in Ishida et al. [1981], Goto [1992] and Williams et al.[2001].

6.5.1 Conceptual Choices for Averaged Results

In Section 6.4, all the results for performance change were expressed on a mass-average basis. However, the impeller is not an isolated device and we have to consider the results in the context of a downstream vaned diffuser (the original pump has a 9-blade vaned diffuser).

Philips [1997], Filipenco et al. [2000] and Deniz et al. [2000] showed that vaned diffuser channel performance is weakly dependent on inlet blockage, but strongly affected by momentum-averaged flow angle. The three authors used the following definition for the momentum-averaged flow angle:

$$\overline{\alpha} = \arctan\left(\frac{\overline{V_\theta}^M}{\overline{V_r}^M}\right). \qquad (6.5\text{-}1)$$

The CFD results show that tip leakage flow generates blockage and modifies the flow angle as defined in Equation 6.5-1. Hence, we can expect the diffuser static pressure recovery to be affected by the flow non-uniformity at impeller exit. Filipenco et al. [2000] and Deniz et al. [2000] showed that mass-averaged stagnation pressure at diffuser inlet is satisfactory in correlating the diffuser pressure recovery. Based on the above considerations, it is appropriate to express the results on a mass average basis.

---

[1] In Section 6.5, the word "model" refers to the flow model developed in Section 6.4 as summed-up in Figure 6.4-3.

6.5.2 Flow Model Results

The model described in Section 6.4 has been applied to two impeller designs, the baseline pump with flow acceleration within the impeller, and the modification with diffusion. The blockage model has therefore been tested both in a favorable and an adverse reduced static pressure gradient. The inputs (blade loading, velocity field, performance measure) were derived from the CFD results.

This section presents a comparison between CFD results and model estimates for both impeller designs.

6.5.2.1 Pressure Rise Sensitivity

Table 6.5-1 summarizes the results for the baseline impeller design for both static and stagnation pressure rise coefficient. All the numerical results shown in Table 6.5-1 are sensitivity parameters. The difference between model estimation and computation results is shown on the right column. The mixing (and blockage) sensitivity refers to the percent change in pressure rise for each percent of tip clearance ratio increase due to mixing loss (and blockage effect). The "Total" sensitivity column gives the percent change in pressure rise for each percent of tip clearance ratio increase, combining the effects of mixing and blockage.

Stagnation Pressure Rise Coefficient Sensitivity					
		Mixing	Blockage	Total	Difference (%)
Baseline Impeller	Model	0.75	1.58	2.34	21
	CFD			2.98	
Modified Impeller	Model	1.13	0.75	1.88	19
	CFD			2.32	
Static Pressure Rise Coefficient Sensitivity					
		Mixing	Blockage (2D wake model)	Total	Difference (%)
Baseline Impeller	Model	0.95	1.3	2.26	15
	CFD			2.65	
Modified Impeller	Model	1.82	0.42	2.24	11
	CFD			2.01	

Table 6.5-1: Baseline and Modified Impellers: Model VS. Computations for Pressure Rise Coefficient Sensitivity to Tip Clearance Variation

The model captures correctly the computed trends based on CFD: there is a maximum 21% difference interval between model estimates (shown in red, total sensitivity column) and CFD results (shown in dark green in the "Total" sensitivity column). Blockage appears to be the dominant cause of performance decrease for the baseline impeller (blockage sensitivity column versus mixing sensitivity column for the baseline impeller), while mixing loss dominates for the redesigned one (blockage sensitivity column versus mixing sensitivity column for the baseline impeller). We will discuss these trends in the parametric study of Section 6.6.

6.5.2.2 Efficiency Sensitivity

From the pressure rise across the impeller and shaft work, we can derive the efficiency as

$$\eta = \frac{CP_T}{H_{Euler}} = 1 - \frac{P_{loss}}{H_{Euler}}. \qquad (6.5\text{-}2)$$

Hence, the variation of efficiency with tip clearance is given by

$$\Delta\eta = \frac{-P_{loss}}{H_{Euler}} + \frac{P_{loss}\Delta H_{Euler}}{H_{Euler}^{2}}. \qquad (6.5\text{-}3)$$

From Section 6.4, we know that $\Delta CP_{T,blockage}$ corresponds to the Euler work variation with tip clearance (work input decrease) and therefore

$$\Delta\eta = \frac{-P_{loss}}{H_{Euler}} + (1-\eta_0)\frac{\Delta CP_{T,blockage}}{H_{Euler}} \qquad (6.5\text{-}4)$$

where η_0 is the impeller efficiency for no tip clearance.

Table 6.5-2 summarizes the results for efficiency sensitivity obtained using the model and from the computation at design point. The baseline impeller is more sensitive than the impellers in Table 6.1-1 in terms of pressure rise but its efficiency sensitivity is comparable to those impellers. The model captures both trends, and the results can be explained in terms of Equation 6.5-4. While blockage is the leading term in pressure sensitivity, its effect on efficiency is less (multiplied by the factor $(1-\eta_0)$ in Equation 6.5-4), and the efficiency drop is mainly due to mixing loss.

Efficiency Sensitivity	Model	CFD	Difference (%)
Baseline Impeller	0.8	0.6	33
Modified Impeller	1.13	0.84	34

Table 6.5-2: Baseline and Modified Impellers: Model VS. Computations for Efficiency Sensitivity to Tip Clearance Variation

6.5.2.3 Blade Loading

The average blade loading (chordwise average pressure difference across the blade) can be related to the work input as

$$\frac{\overline{\Delta P}}{\rho} = \frac{2\pi Q}{Z\int_{r_1}^{r_2} A(r)dr}\left(r_2\overline{V_{\theta 2}}^{M} - r_1\overline{V_{\theta 1}}^{M}\right), \qquad (6.5\text{-}5)$$

where Q is the impeller flow rate, $A(r) = 2\pi hr$ is the area at radius r with h the blade span, and V_θ is the absolute tangential velocity.

We can simplify this expression for the baseline impeller, using its area distribution, to obtain the following expression for the change in blade loading due to blockage effects on stagnation pressure rise:

$$\Delta\overline{CP}_{S,loading} = \frac{2\pi\phi_2}{Z\left(1-\frac{r_1}{r_2}\right)}\Delta CP_{T,blockage}. \qquad (6.5\text{-}6)$$

The average loading sensitivity to tip clearance variation obtained from CFD results is compared with the model result using Equation 6.5-5 in Table 6.5-3.

	Model	CFD	Difference (%)
Loading Drop	1.52	1.60	5

Table 6.5-3: Baseline Impeller: Model VS. Computations for Loading Sensitivity due to Tip Clearance Variation

6.5.2.4 Summary of Sensitivity Estimates

The results from Tables 6.5-1 and 6.5-2 are summarized in Figure 6.5.1 to indicate the comparison of CFD and model results. If the model perfectly captured the computed results, the bars in Figure 6.5-1 would lie on the green line. The vertical gap between the bars and the green line is the difference between computation and model results, expressed in percent of the CFD results. Static and stagnation pressure sensitivities are within 20%. Efficiency sensitivity is overestimated by about 35% in both cases.

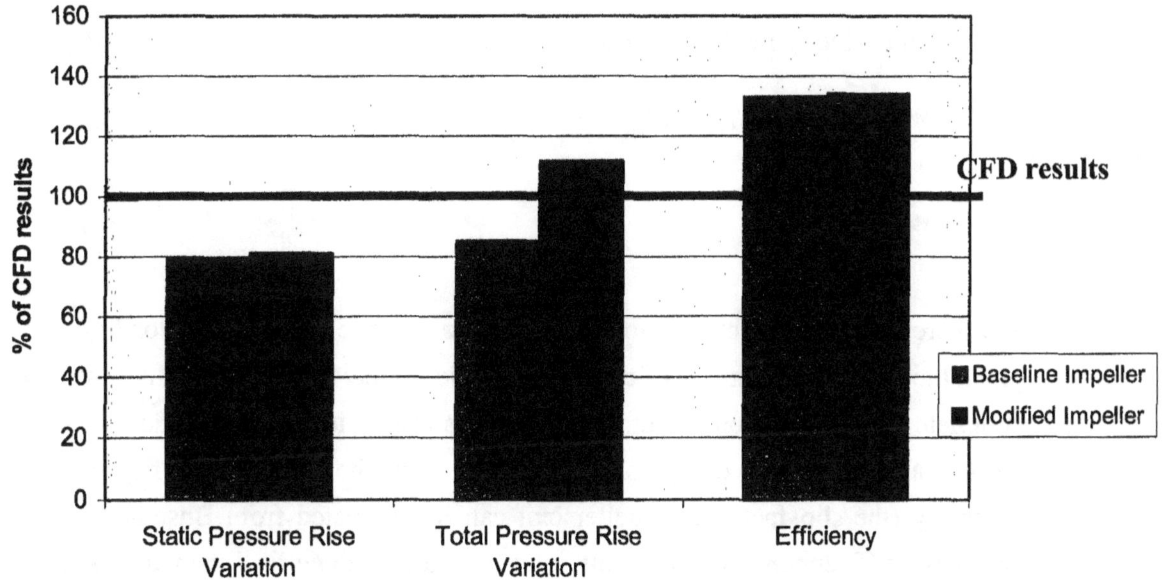

Figure 6.5-1: Model Assessment

6.5.3 Model Input

We have used CFD results to obtain model inputs in order to assess the adequacy of the model. To use the flow model as a predictive tool, however, we need to define what information about a given impeller is required and what approximations need to be made. These aspects are examined next.

6.5.3.1 Description of Input

We distinguish between two types of input:

- General information about the impeller:
    - Geometry: inlet and outlet blade angle, radius and span, tip clearance configuration, number of blades
    - Design parameters: flow coefficient, rotational speed, static and stagnation pressure rise
- Operating information for the impeller
    - Blade loading distribution
    - Velocity field: more precisely W_s and W_p distributions
    - Reduced static pressure field

While the first group of data needs to be provided, the second group can be estimated as described below.

6.5.3.1.1 Velocity field

The velocity field can be estimated as

$$W_S{}^2 - W_P{}^2 = 2CP_{loading}U_{tip}{}^2 \tag{6.5-7}$$

$$\overline{W} \approx \frac{W_P + W_S}{2}. \tag{6.5-8}$$

We therefore need to obtain information on the average velocity and loading. If we are given the complete geometry of the impeller (meridional and radial profile), we know the complete area distribution. Assuming the flow is well guided by the blades, we can derive the average velocity at design on any surface from the design mass flow-rate and the geometry (the slip factor at impeller exit can be estimated from Busemann or Stodola correlations [Brennen, 1994]). If only inlet and outlet geometric parameters (span, radius, blade angle) are known, a "mean" area distribution needs to be interpolated.

6.5.3.1.2 Loading

If the impeller loading curve (blade loading versus radius) is not known, an average value can be derived using the following approximation

$$\overline{CP}_{S,loading} = \frac{2\pi Q_{total}}{Z\Omega \int_{r_1}^{r_2} A(r)dr} CP_T. \tag{6.5-9}$$

Assuming loading is constant along the impeller (a simple approximation which is supported by the CFD results for the baseline impeller), Equation 6.4-13 for mixing loss reduces to

$$P_{loss} = \frac{Z}{4\pi r_2 \varphi_2}\left(2\overline{C}_{P,loading}\right)^{3/2} C_d \lambda \int_{r_1}^{r_2} \frac{dr}{\sin(\beta_b)}. \tag{6.5-10}$$

6.5.3.1.3 Reduced Static Pressure Gradient

With no streamwise gradient of reduced stagnation pressure, which is the case in the absence of viscous effects, the reduced static pressure field can be deduced from the velocity field:

$$CP_{red} = \frac{\Delta P_{red}}{\frac{1}{2}\rho W_1{}^2} = \frac{W_1{}^2 - W_2{}^2}{W_1{}^2}. \tag{6.5-11}$$

Indeed, the model input is based on the shrouded case, and efficiency is high in centrifugal pumps (more than 90% for the baseline and modified impellers), so the approximation in Equation 5-11 seems reasonable.

A summary of the model input and application is given in Figure 6.5-2.

6.5.3.2 Example

In this section, we compare the model results (stagnation pressure rise, baseline and modified impellers) for two situations, one with fully known inputs (CFD) and one with estimated inputs (Equations 6.5-7, 6.5-8, 6.5-9 and 6.5-11).

Table 6.5-4 shows the comparison between the average loading (CFD results) and the estimate provided by Equation 6.5-9, for the baseline and the modified impeller.

	Blade Loading	
	Baseline Impeller	Modified Impeller
From CFD	0.073	0.110
Equation 5-9	0.079	0.111
difference (%)	8.22	0.74

Table 6.5-4: Baseline and Modified Impeller: Loading Approximation

Table 6.5-5 compares the sensitivity of the stagnation pressure rise due to mixing using CFD or approximate inputs to the models, for both the baseline impeller and the modified impeller. There is respectively a 16% and 2% difference between the model estimates based on the full and approximate input.

	Sensitivity of Stagnation Pressure Rise Coefficient due to Mixing	
	Baseline Impeller	Modified Impeller
CFD Input	0.75	1.13
Approxiate Input	0.87	1.11
Difference (%)	16	2

Table 6.5-5: Baseline Pump: Comparison on Mixing Impact

In summary, the use of approximate input to the model has been assessed for two examples and found to agree within a 16% difference interval.

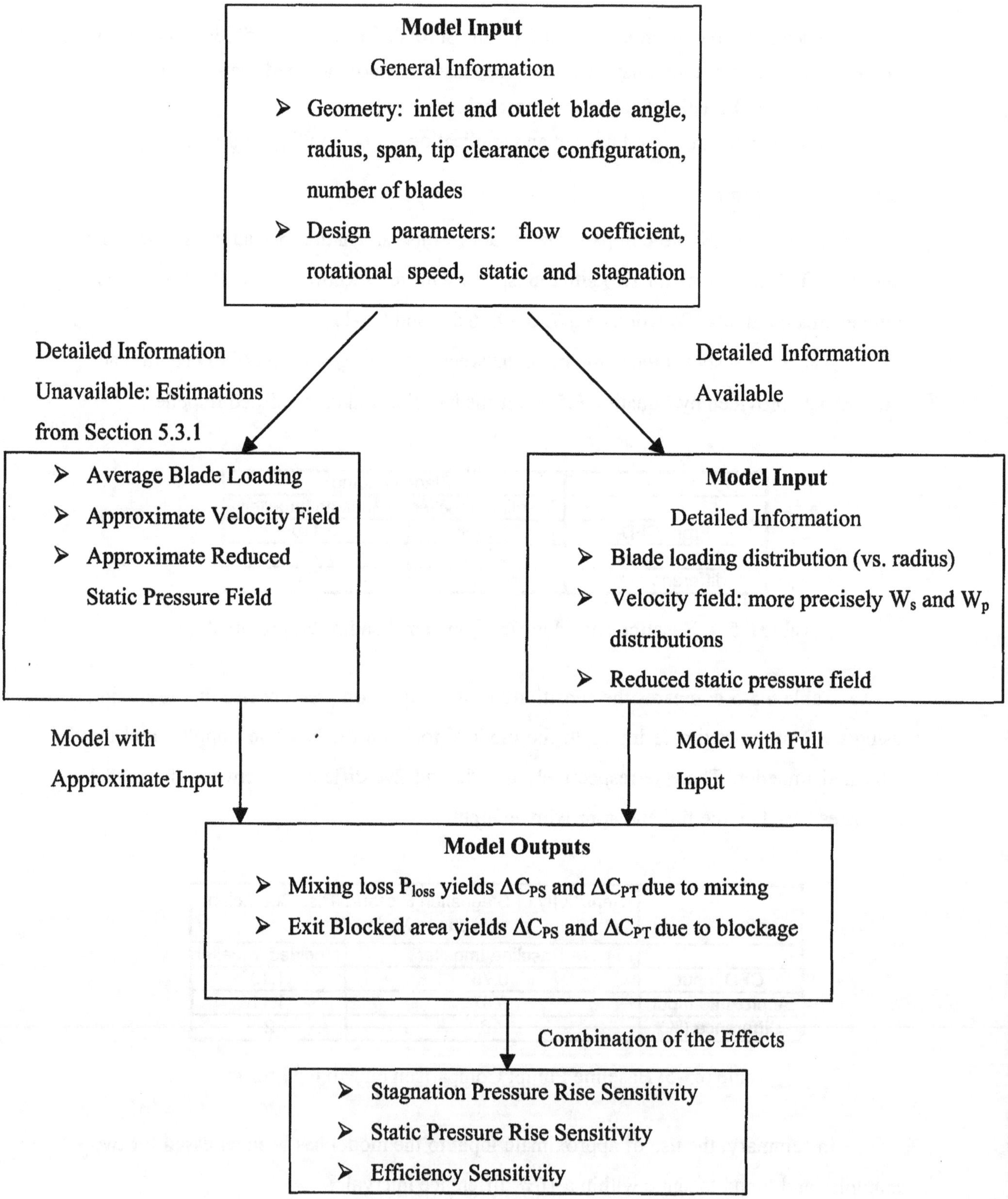

Fig.6.5-2: Summary on Model Input

6.5.4 Results with Impellers of Different Design

The model is now applied to three reported impeller designs that were found to provide sufficient information: [Ishida et al., 1981], [Goto, 1992] and [Williams et al., 2001]. Overall design and geometrical information were given so the approximate expressions for velocity and loading derived in Section 6.5.3.1 were used.

Table 6.5-6 shows the results for three different impellers, model *versus* reported results. There is some uncertainty concerning the published results. For the Rocketdyne impeller, two different sensitivities were obtained from CFD and experimental studies (the lowest estimate is from CFD). In the other cases, the sensitivities were measured from the given performance graphs and there may be up to 10% error associated with this. Nevertheless the model estimates are within 40% of the published results [Ishida et al., 1981], [Goto, 1992] and [Williams et al., 2001]. Two of the three compressors exhibit lower sensitivity than the pump studied here and the model captures this trend.

	Williams et al. [8]	Goto [7]	Ishida, Senoo [4]
Model	0.8	2.2	1.4
Reported Data	0.8-1.1	2.5	1.0

Table 6.5-6: Model VS. Reported Results on Stagnation Pressure Rise Coefficient Sensitivity to Tip Clearance Variation

6.5.5 Summary

This section described application of the flow model developed in Section 6.4 to five impeller pump designs. The model was assessed against the computed results for the baseline and modified impellers presented in Section 6.3. Its applicability was also examined using impellers from Ishida et al. [1981], Goto [1992] and Williams et al.[2001]. With full input, the model captured the performance sensitivity within 20% difference interval of CFD results. With inputs based on the approximations developed in Section 6.5.3.1, the model was within 40% of the reported data. The model also enabled one to distinguish between mixing and blockage contributions to performance decrease.

6.6 Parameterization of Performance Sensitivity to Tip Clearance Variation

6.6.1 Introduction

The model developed in Section 6.4 implies that flow features associated with tip clearance are generally similar in low specific speed centrifugal and high specific speed axial turbomachines. The difference lies in the degree of contribution of different physical effects to the change in performance. The objective of this section is to identify the parametric dependence of performance sensitivity and to define the relative contributions from mixing and from tip clearance-related blockage. Design guidelines are also suggested to mitigate the sensitivity of impeller performance to tip clearance variation under engineering constraints delineated in Section 6.6.2.1.2.

6.6.2 Expressions for the Stagnation Pressure Rise and Efficiency Sensitivity: Governing Parameters

6.6.2.1 Stagnation Pressure Rise Coefficient Sensitivity

6.6.2.1.1 Blockage

As shown in Section 6.4, three contributions to the decrease in pressure rise coefficient due to blockage can be identified.

From Equation 6.4-27, we have

$$\Delta CP_{T,blockage} = -\overline{\tan\beta_2}^{M}\,\frac{\overline{\Delta W_2}^{M}}{\overline{W_2}^{M}}\,\overline{\phi_2}^{M}. \qquad (6.4\text{-}27)$$

The contributions are thus

- Average exit relative flow angle, β_2
- Exit flow coefficient
- Normalized change in exit velocity magnitude between shrouded case and tip clearance case which is set by the exit blockage

We wish to derive expressions for the sensitivity of the pressure rise coefficient to tip clearance variation. For blocked areas which are small relative to the geometric exit area, Equation 6.4-28 for the velocity change (between shrouded and tip clearance cases) can be simplified to

$$\frac{\overline{\Delta W_2}^M}{\overline{W_2}^M} \approx \frac{A_{b,r2}}{A_{r2}}. \quad (6.6\text{-}1)$$

Because the exit blockage is proportional to tip clearance ratio (from the wake model),

$$\frac{\overline{\Delta W_2}^M}{\overline{W_2}^M} \approx k\lambda. \quad (6.6\text{-}2)$$

In Equation 6.6-2, k represents the percent exit area blocked for each percent of tip clearance ratio increase and is a function of blade loading, reduced pressure gradient and blade geometry. Using this parameter and Equations 6.4-23 and 6.4-27, an approximate expression for stagnation pressure rise sensitivity due to blockage as

$$s_{T,blockage} = k\left(\frac{1}{1-\overline{\phi_2}^M \overline{\tan\beta_2}^M} - 1\right). \quad (6.6\text{-}3)$$

The three contributions of Equation 6.4-27 appear in Equation 6.6-3 with similar trends. The sensitivity increases with increasing exit mass averaged flow coefficient and with increasing exit relative flow angle, and the sensitivity is proportional to k.

The parameter k can be (numerically) obtained by applying the wake model to the discrete blade segments. This requires one to determine the velocity distribution in the impeller. The value of k increases with:

- Adverse reduced pressure gradient, which exacerbates the flow non-uniformity
- Blade loading, which sets the initial velocity defect ratio $\gamma_1 = \frac{W_P}{W_S}$ and the initial wake area
- Blade length, which sets the leakage area
- Number of blades, which impacts the blade loading and the slip factor.

6.6.2.1.2 Constraints and Requirements

In this section, we define some design guidelines for impellers that are less sensitive to tip clearance. This is done under the following constraints.

1. Fixed diameter

The impellers should to be as compact as possible, to limit its weight as well as its extent.

2. Fixed rotational speed

Stress requirements as well as temperature constraints limit the rotational speed of the impeller.

3. Fixed stagnation pressure

The pump is supposed to pressurize the fluid at a level that corresponds to the combustion chamber requirements. Hence, the stagnation pressure rise in the impeller (which sets the achievable static pressure rise of the pump stage) is a constraint. Since the rotational speed and the diameter are fixed, this means that the stagnation pressure coefficient is kept constant. As noted in Section 6.4 (see Equations 6.4-23 and 6.4-24),

$$CP_T \approx 1 - \overline{\phi_2}^M \, \overline{\tan\beta_2}^M . \qquad (6.6\text{-}4)$$

We therefore require

$$\overline{\phi_2}^M \, \overline{\tan\beta_2}^M = \text{constant} = a . \qquad (6.6\text{-}5)$$

4. Mass flow rate

The mass flow rate is set-up by operating thrust requirements and also kept constant.

Efficiency is also an important parameter in pump impeller design but we are not evaluating the change in efficiency induced by a change in design parameters.

6.6.2.1.3 Change in Design Parameters

We next proceed to analyze the impact of change in these design parameters under the conditions of Section 6.6.2.1.2. Combining Equations 6.6-3 and 6.6-5, we obtain the result that $s_{T,blockage}$ is only affected by a change in k:

$$s_{T,blockage} = \frac{ka}{1-a} . \qquad (6.6\text{-}6)$$

We can illustrate this situation by considering the effect of a change in blade exit angle. An increase in blade exit angle must be accompanied by a flow coefficient decrease. This results in a blade loading decrease (see Equation 6.6-7), which tends to lower k (decrease in the initial wake defect ratio, better alignment of the tip vortex trajectory and the mainstream trajectory). From Equation 6.5-9, the average loading can be approximated by

$$\overline{CP}_{loading} = \frac{2\pi}{Z} \phi_2 \frac{1}{\int_{r_1}^{r_2} \frac{A(r)dr}{A(r_2) r_2}} CP_T . \qquad (6.6\text{-}7)$$

Increasing the blade exit angle leads to a blade length increase, which tends to increase k. The trade-off between these two competitive effects on k cannot be obtained analytically, but implementing the model on the baseline pump with varying blade exit angle, we find that k increases with blade exit angle increase[1]. This is shown in Figure 6.6-1.

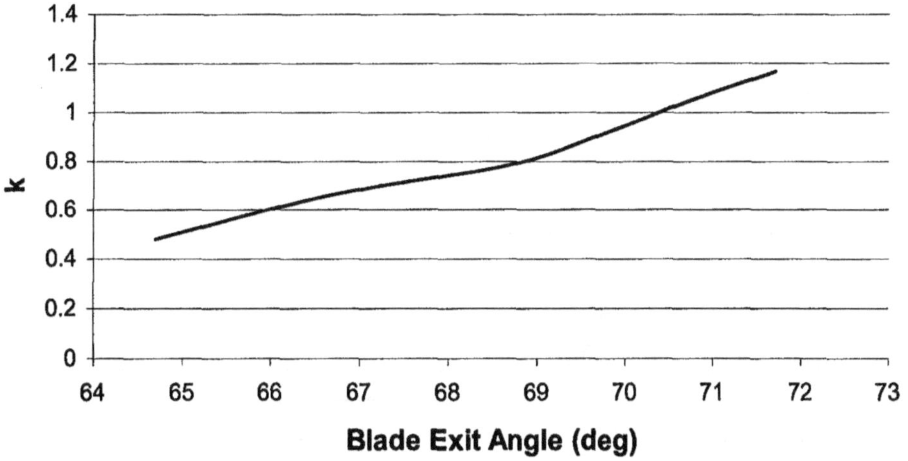

Fig. 6.6-1: Effect of a Change in Blade Exit Angle on the Parameter *k* for the Baseline Impeller

The same analysis is undertaken for a change in flow coefficient, and the results are shown in Figure 6.6-2 in which *k* is seen to decrease with increasing flow coefficient.

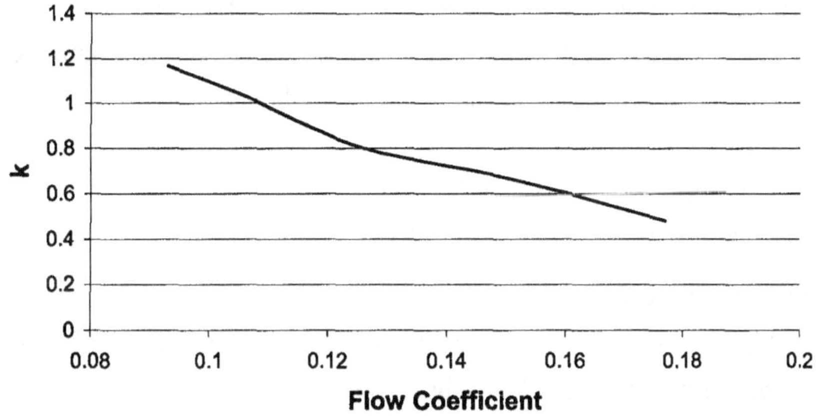

Fig. 6.6-2: Effect of a Change in Flow Coefficient on the Parameter *k* for the Baseline Impeller

Finally, we show in Figure 6.6-3 that *k* decreases with an increase in the number of blades. On the one hand, increasing the number of blades decreases the blade loading

[1] We approximated the deviation angle due to slip as a constant on the considered interval of blade angles.

and reduces the slip factor (Stodola correlation, [Brennen, 1994]), which tends to lower *k*. On the other hand, it reduces the blade to blade passage area at impeller exit, which tends to increase *k*. The observed numerical trade-off between these two effects is a decrease in k.

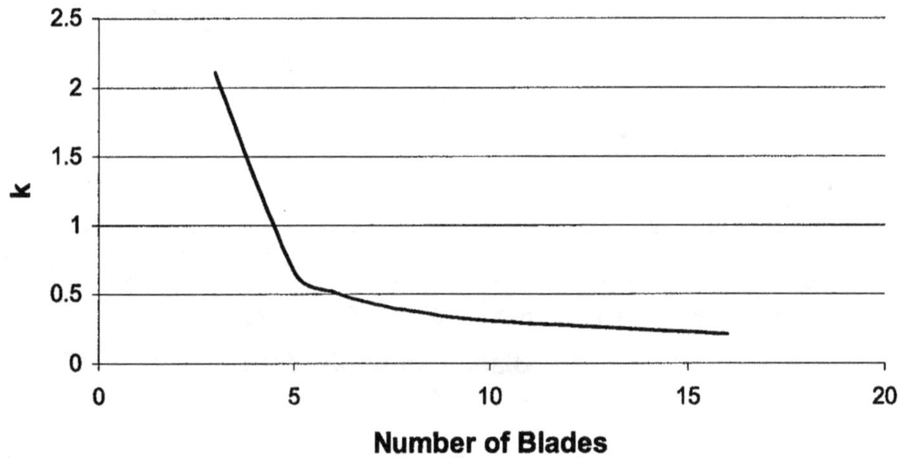

Fig. 6.6-3: Effect of a Change in Blade Number on the Parameter *k* for the Baseline Impeller

6.6.2.1.4 Mixing

The mixing loss has a similar effect on static and stagnation pressure rise coefficient. To isolate the contribution to mixing loss, we use the approximate expression derived in Section 6.5 assuming constant blade loading along the impeller. The loss parameter, P_{loss}, is

$$P_{loss} = \frac{Z}{4\pi r_2 \phi_2}\left(2\overline{CP}_{loading}\right)^{3/2} C_d \lambda \int_{r_1}^{r_2} \frac{dr}{\sin(\beta_b)}. \qquad (6.5\text{-}10)$$

There are four different contributions. The first is geometric and incorporates number of blades, blade length (integral term) and exit radius. P_{loss} increases with blade length, which is a function of blade angle. The second is blade loading, the third is flow coefficient, and the fourth is the contraction coefficient (vena contracta).

The loading is linked to geometry as well as flow coefficient. We therefore take these links into account to simplify 6.5-10 so as to elucidate the various independent contributions. We rearrange Equation 6.5-10 so that the flow coefficient appears as the main variable. Thus substituting Equations 6.6-1 and 6.6-6 into Equation 6.5-10 yields the following expression for P_{loss} as

$$P_{loss} = \sqrt{\frac{4\pi\overline{\phi_2}^M}{Z}} \left(\frac{1-\overline{\tan\beta_2}^M \overline{\phi_2}^M}{\int_{r_1}^{r_2} \frac{A(r)dr}{A(r_2)r_2}} \right)^{3/2} C_d \lambda \int_{r_1}^{r_2} \frac{dr}{r_2 \sin(\beta_b)}. \quad (6.6\text{-}8)$$

We can normalize the expression by dividing through by $\lambda(CP_T)_0$ (see Section 6.1.2.1) to obtain the sensitivity as

$$s_{T,mixing} = \sqrt{\frac{4\pi\overline{\phi_2}^M (1-\overline{\tan\beta_2}^M \overline{\phi_2}^M)}{Z}} \left(\int_{r_1}^{r_2} \frac{A(r)dr}{A(r_2)r_2} \right)^{-3/2} C_d \int_{r_1}^{r_2} \frac{dr}{r_2 \sin(\beta_b)}. \quad (6.6\text{-}9)$$

Equation 6.6-9 shows the part of the stagnation pressure rise sensitivity, which is due to mixing as a function of geometric parameters and flow coefficient.

We first address the response of a given impeller to tip clearance variation at off-design flow rate; $s_{T,mixing}$ is a function of the flow coefficient alone (Equation 6.6-9), and has a maximum at $\phi_0 = \frac{1}{2\tan\overline{\beta_2}^M}$.

We now analyze the effects of a change in design parameters under the conditions listed in Section 6.6.2.1.2.

1. Change in Blade Exit Angle

For a blade profile that is logarithmic, the blade exit angle is given by

$$\beta_b = \frac{\pi}{2} - \beta_2. \quad (6.6\text{-}10)$$

Using Equation 6.6-5 in Equation 6-9 yields

$$s_{T,mixing} \propto \sqrt{\frac{1}{\overline{\sin(2\beta_2)}^M}}. \quad (6.6\text{-}11)$$

The sensitivity is smallest for β_2 equal to 45 degrees and increases with increasing exit relative flow angle for $\beta_2 \geq 45\deg$, which is representative of centrifugal pumps. This trend reflects the trade-off between beneficial effects of backsweep reduction (blade length decrease) and detrimental effects of backsweep reduction (increase in the flow coefficient and blade loading).

2. Change in Flow Coefficient

Substituting Equation 6.6-5 in Equation 6.6-9, we obtain

$$s_{T,mixing} \propto \sqrt{\overline{\phi_2}^M \left(\left(\frac{a}{\overline{\phi_2}^M} \right)^2 + 1 \right)}. \qquad (6.6\text{-}12)$$

Equation 6.6-12 shows the sensitivity $s_{T,mixing}$ is a minimum when $\overline{\phi_2}^M = a = 1 - CP_t$, which implies $\beta_2 = 45\deg$ (See Equation 6.6-5). If $\beta_2 > 45\deg$, $\overline{\phi_2}^M < a$ (see Equation 6.6-5) and the sensitivity decreases when flow coefficient increases (see Equation 6.6-12)

3. Number of Blades

From Equation 6.6-9, $s_{T,mixing}$ decreases with increasing number of blades, with a power (–1/2) dependency.

6.6.2.1.5 Summary

For the baseline impeller, from Figures 6.6-1,6.6-2 and 6.6-3, and the conclusions of Section 6.6.2.1.4, the overall stagnation pressure rise sensitivity (effects of both mixing and blockage) increases with blade exit angle, and decreases with flow coefficient and number of blades.

6.6.2.2 Efficiency Sensitivity

6.6.2.2.1 Analytical Expression

In this section we derive an analytical expression for the efficiency decrease due to tip clearance increase. The efficiency change with tip clearance is given by

$$\Delta\eta = \frac{-P_{loss}}{H_{Euler}} + (1-\eta_0)\frac{\Delta CP_{T,blockage}}{H_{Euler}}. \qquad (6.5\text{-}4)$$

If efficiency is close to unity, which is a reasonable approximation for centrifugal impellers (centrifugal pumps commonly have efficiency higher than 90%), the blockage term is small relative to the mixing term. Thus

$$\Delta\eta \approx \frac{-P_{loss}}{H_{Euler}}. \qquad (6.6\text{-}13)$$

Hence, combining Equations 6.6-13 and the definition of $s_{T,mixing}$ yields

$$s_{T,mixing} = \frac{P_{loss}}{C_{PT}\Delta\lambda} = \frac{\Delta\eta}{\eta\Delta\lambda} = s_{\eta}. \quad (6.6\text{-}14)$$

Efficiency sensitivity is then given by the expression in Equation 6.6-9 and the pump is most sensitive, in terms of efficiency, at $\phi_0 = \frac{1}{2\tan\overline{\beta_2}^M}$.

For the baseline impeller, the maximum sensitivity is roughly $\phi_0 = 0.115$. This is confirmed by CFD results as shown in Figure 6.6-4 which gives a comparison of CFD results and estimation from Equation 6.6-14. The approximate analysis isolates the maximum sensitivity close to design flow coefficients ($\phi = 0.12$) but does not correctly capture the important drop in efficiency sensitivity at flow coefficient away from design.

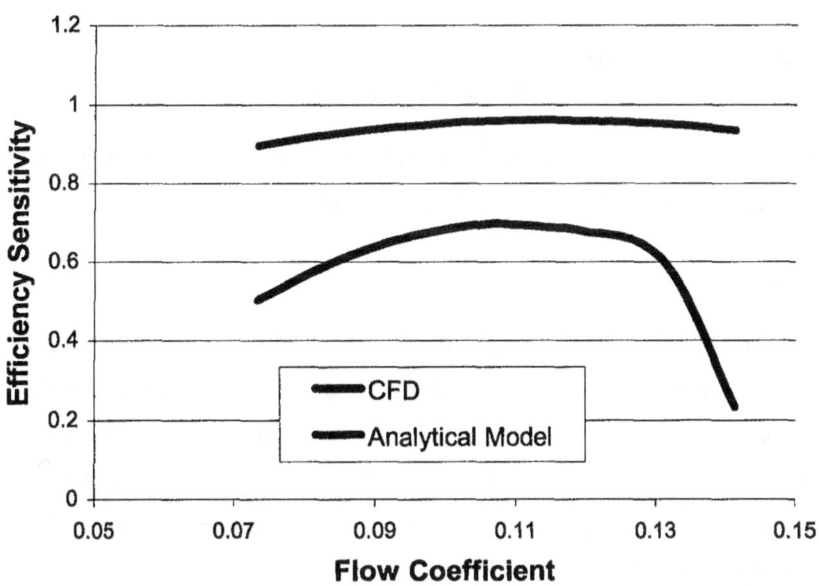

Fig. 6.6-4: Efficiency Sensitivity: Comparison between CFD Results and Results from Analysis for the Baseline Impeller

Figure 6.6-5 shows the same comparison for the modified impeller[2]. It can also be seen that the approximate analytical expression captures the increase in efficiency

---

[2] We did not run calculations at flow coefficients higher than reported in Figure 6.6-5 because the angle mismatching between the inlet flow and the blade angle was too important, leading to unreasonable flow separation at impeller inlet.

sensitivity with flow coefficient. In both cases (Figure 6.6-4 and Figure 6.6-5), the analytical expression yields an estimate for the computed efficiency variation within 30% at design conditions, and the approximate analysis gives the trends of efficiency sensitivity evolution at off-design conditions.

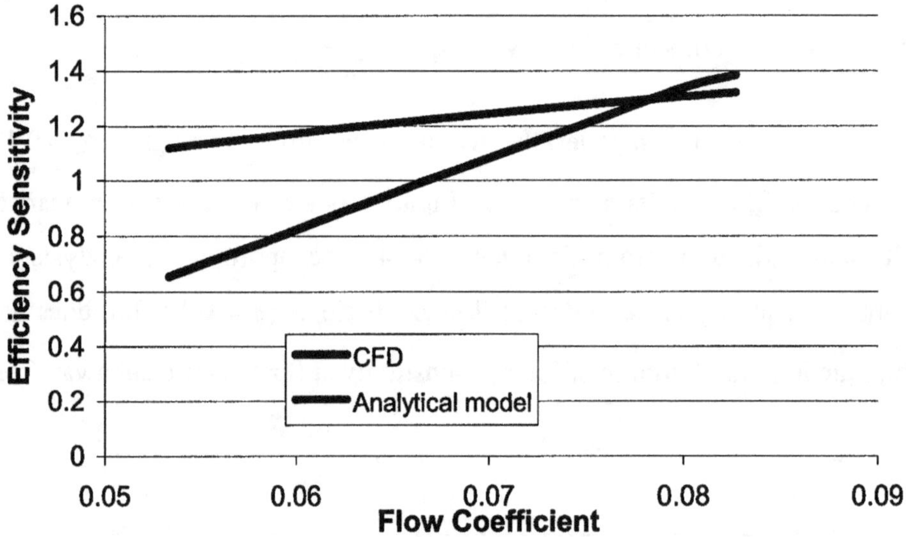

Fig. 6.6-5: Efficiency Sensitivity: Comparison between CFD Results and Results from Analysis for the Modified Impeller

6.6.2.2.2 Summary

The expression for efficiency sensitivity is similar to that for stagnation pressure rise sensitivity due to mixing, and we can apply the trends of Section 6.6.2.1.4 to efficiency sensitivity. Efficiency sensitivity can be reduced by reducing blade exit angle or increasing flow coefficient as long as $\beta_2 \geq 45\,\text{deg}$ (and vice-versa) and by increasing the number of blades. The dependence of sensitivity upon blade number follows that derived by Lauer and al. [1997].

6.6.3 The Relative Importance of Blockage and Mixing

As stated in Section 6.5, the impeller exit blockage can impact diffuser performance, while the consequences of mixing within the impeller are confined to the impeller. It is thus important to distinguish between blockage and mixing, and the flow model developed in Section 6.4 provides this distinction. Figure 6.6-6 shows the relative contribution to performance sensitivity from mixing and blockage effects for the baseline and the redesigned impeller.

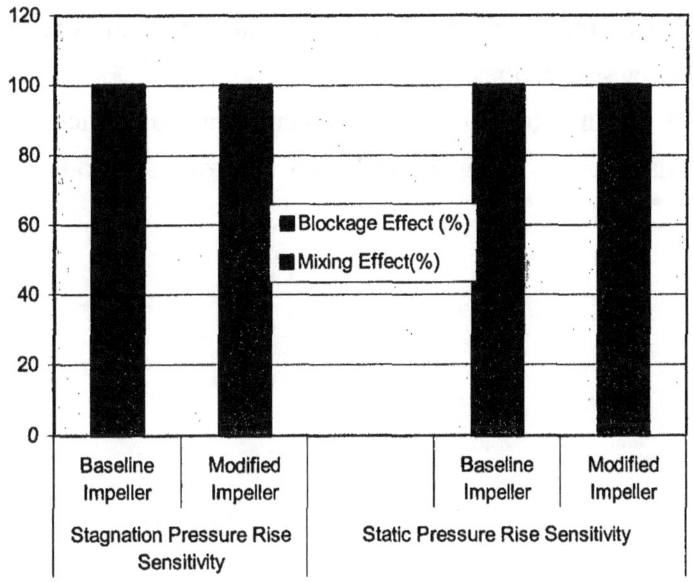

Fig. 6.6-6: Relative Importance of Blockage and Mixing

In the baseline impeller, blockage is the driver for static and stagnation pressure rise sensitivity. In the modified impeller, the relative importance of blockage is diminished and mixing is most important for static and stagnation pressure rise sensitivity

One might think that since the redesigned impeller presents an adverse pressure gradient, the relative importance of blockage should be higher than in the baseline impeller. This is actually not the case and this situation underlines the complexity of trade-offs for blockage evolution. Using Equation 6.6-3, it is seen that the redesigned pump presents a larger proportion of exit blocked area (k increases) but the flow coefficient and exit flow angle are lower and there is a reduction in stagnation pressure rise sensitivity to blockage.

For mixing, the flow coefficient is lower and the blade loading is higher for the modified impeller. As shown in Equation 6.6-9, this leads to higher performance sensitivity.

6.6.4 Summary

Using the flow model presented in Section 6.4, approximate expressions have been obtained for performance sensitivity to tip clearance variation. From these the independent parameters that characterize the performance sensitivity are the blade exit angle, the number of blades and the exit flow coefficient.

The following parametric trends were found:

- Low blade exit angle and high flow coefficient reduce the stagnation pressure rise and efficiency sensitivities.
- Increasing the number of blades is beneficial for efficiency and stagnation pressure rise sensitivity, in accord with the results reported by Lauer and al. [1997].

6.7 Summary and Conclusions

6.7.1 Summary

The effects of tip clearance variation on centrifugal impeller performance have been investigated using three-dimensional CFD simulations. Two impeller geometries were used: a baseline configuration having flow acceleration and a modified configuration with diffusion, representative of rocket motor turbomachinery.

1. The impact of tip clearance on flow structure was examined from visualization of the secondary flow patterns in the impeller passage, from leading edge plane to trailing edge plane. Contours of reduced stagnation pressure, or relative velocity magnitude, were also used to isolate the high loss and low velocity regions.
2. A tip vortex trajectory analysis developed for axial compressors [Chen etal., 1991] was adapted to centrifugal impellers to estimate the path of tip leakage streamtubes.
3. Based on the above results, an explanatory model for tip clearance induced performance decrease has been developed. The model, which has been assessed against CFD results, provides a tool for describing the impeller performance sensitivity (decrease in static pressure rise, stagnation pressure rise and efficiency). The input parameters (geometric data, velocity field, blade loading) have also been defined and a method presented for estimation of these if full information is not provided.
4. An approximate analytical expression was derived for the effect of mixing loss and efficiency sensitivities. The results from this expression were assessed against the CFD results and shown to describe the trend of efficiency sensitivity at off-design conditions for the baseline and the modified impeller.
5. The model was used to define the parametric dependence of impeller sensitivity to tip clearance variation.

6.7.2 Conclusions

1. The secondary flow structure in an unshrouded impeller is dominated by the tip vortex, the size of which increases with tip clearance ratio and streamwise distance. The correspondence between the tip vortex trajectory and the low reduced stagnation pressure regions indicates that the tip clearance related velocity defect region is embedded in the clearance vortex.
2. The flow model developed shows two specific aspects of tip leakage flow which affect performance. One is mixing (approximated by loss of tip leakage cross-flow

kinetic energy) and the other is blockage (evolution of the low streamwise velocity tip clearance flow).

3. The flow model was assessed against several centrifugal pumps and one centrifugal compressor and found to approximately describe the development of tip leakage flow.
4. The parameterization of tip clearance related performance decrease shows that the efficiency sensitivity increases with increasing blade exit angle, decreasing flow coefficient and decreasing number of blades.

6.8 Nomenclature

A	Area
c	chord
C_d	Contraction factor
CP_S	Static pressure rise coefficient, $\dfrac{\overline{P_S}^M - \overline{P_{S,inlet}}^M}{\rho U_{tip}^2}$
CP_T	Stagnation pressure rise coefficient, $\dfrac{\overline{P_T}^M - \overline{P_{T,inlet}}^M}{\rho U_{tip}^2}$
$CP_{T,red}$	Reduced stagnation pressure coefficient, $\dfrac{P_{T,red} - \overline{P_{T,red,inlet}}^M}{\rho U_{tip}^2}$
D	Relative velocity diffusion ratio
h	Blade height
H_{Euler}	Work input Coefficient, $H_{Euler} = \dfrac{W_{shaft}}{\rho U_{tip}^2}$
l	Streamwise length
$\dot{m}$	Mass flow rate
n	blade normal
r	Radius
P_S	Static pressure
P_T	Total pressure
P_{red}	Reduced pressure
$P_{T,red}$	Reduced Total Pressure
Q	flow rate
s	Entropy
s_X	Sensitivity of the parameter X
T	Temperature
U	Impeller rotational speed
V	Absolute velocity
w	Defect relative velocity
W	Relative velocity
W_{shaft}	Shaft Work
x	axial coordinate
y_c	Distance from the tip vortex center to the blade
Z	Number of blades
z_c	Distance from the tip vortex center to the casing
α	Angle between tip clearance flow and blade normal
β	Relative flow angle
β_b	Blade angle
Δ	Variation symbol
ϕ	Flow coefficient
γ_1	Wake defect ratio

λ	Tip clearance ratio
η	Efficiency
θ	plane change angle
Π	Shaft Power
ρ	Flow density
τ	Tip clearance gap height
χ	Inlet to outlet radius ratio $\chi = \frac{r_1}{r_2}$
ω	vorticity

Subscripts

b	Blockage
b	Blade
c	Clearance
e	Exit
f	final
n	Normal
m	Main flow
p	Pressure side
r	Radial
s	Suction side
t	Tangential
sf	Secondary flow
x	Axial
1	Impeller inlet
2	Impeller exit

Operators

$\overline{}^{M}$	Mass average
$\overline{}^{A}$	Area average
$\underline{u}$, $\mathbf{u}$	Vector
$\cdot$	Scalar product

Acronyms

LE	Leading edge
PS	Pressure side
SS	Suction side
TE	Trailing edge

6.9 References

Brasz, J.J., 1988, "Investigation into the Effect of Tip Clearance on Centrifugal Compressor Performance", ASME Paper 88-GT-190.

Brennen, C.E., 1994, "Hydrodynamics of Pumps", Concepts ETI, Inc, and Oxford University Press.

Chen, G.T., Greitzer, E.M., Tan, C.S., and Marble, F.E., 1991, "Similarity Analysis of Compressor Tip Clearance Flow Structure", *Journal of Turbomachinery*, Vol. 113, pp260-271.

Denton, J.D., 1993 "Loss Mechanisms in Turbomachines", *Journal of Turbomachinery*, Vol. 115, pp621-656

Deniz, S., Greitzer, E.M., Cumpsty N.A., 2000, "Effects of Inlet Flow Field Conditions on the Performance of Centrifugal Compressor Diffusers, Part 2: Straight-Channel Diffuser", *Journal of Turbomachinery*, Vol. 122, pp11-21.

Engeda, A., Strate, W. P., Rautenberg, M., 1988, " Correlation of Tip Clearance Effects to Impeller Geometry and Fluid Dynamics", ASME, 3rd, Gas Turbine Conference, Amsterdam.

Engeda, A., Rautenberg, M., 1987, "Comparison of the Relative Effect of Tip Clearance on Centrifugal Impellers", *Journal of Turbomachinery*, Vol. 109, pp 545-549.

Engeda, A., 1995, " Tip-Clearance Flow Effects on Centrifugal Impellers", ASME, FED-Vol. 217.

Filipenco, V. G., Deniz, S., Johnston, J.M., Greitzer, E.M., Cumpsty. N. A., 2000, "Effects of Inlet Flow Field Conditions on the Performance of Centrifugal Compressor Diffusers, Part 1: Discrete-Passage Diffuser", *Journal of Turbomachinery*, Vol. 122, pp1-10.

Fluent 6.0 User's Guide, Fluent.Inc.

Goto, A., 1992, "Study of Internal Flows in a Mixed-Flow Pump Impeller at Various Tip-Clearances Using 3D Viscous Flow Computations", *Journal of Turbomachinery*, Vol.114, pp373-382.

Hoshide, R.K, and Nielson, C.E,. 1972 "Study of Blade Clearance Effects on Centrifugal Pumps", NASA Paper CR-120815, NASA-Lewis Research Center.

Ishida, M., and Senoo, Y., 1981, "On the Pressure Losses Due to the Tip Clearance of Centrifugal Blowers", *Journal of Engineering for Power*, Vol. 103, pp 271-278.

Ishida, M., Ueki, H., and Senoo, Y., 1990, " Effect of Blade Tip Configuration on Tip Clearance Loss of a Centrifugal Impeller", *Journal of Turbomachinery*, Vol. 112, pp 14-18.

Ishida, M., Senoo, Y., and Ueki, H., 1990, "Secondary Flow due to the Tip Clearance at the Exit of Centrifugal Impellers", *Journal of Turbomachinery*, Vol. 112, pp 19-29.

Khalid, S.A., 1995, "The Effects of Tip Clearance on Axial Compressor Pressure Rise", Ph. D. Thesis, Department of Aeronautics and Astronautics, Massachusetts Institute of Technology.

Khalid, S.A., 1999, Khlasa, A.S., Waitz, I.A., tan, C.S., Greitzer, E.M., Cumpsty, N.A., Adamczyk. J.J., Marble, F.E., "Endwall Blockage in Axial Compressors", *Journal of Turbomachinery*, Vol. 121, pp499-509.

Lauer, J., Ludwig, G., Jaeger, R., and Stoffel, B., 1997 "Tip Clearance Sensitivity of Centrifugal Pumps with Semi-Open Impeller", ASME Fluids Engineering Division Summer Meeting, FEDSM97-3366.

Philips, M.S., 1997, "A Computational Investigation of the Effects of Inlet Flow Conditions on Vaned Diffuser Performance", M.S. thesis, Department of Aeronautics and Astronautics, Massachusetts Institute of Technology

Pruger, G. H., 2001, Rocketdyne, Personal Communication.

Senoo, Y., and Ishida, M., 1985 "Pressure Loss Due to the Tip Clearance of Impeller Blades in Centrifugal and Axial Blowers", ASME Paper 85-GT-196

[14] Senoo, Y., "Mechanics on the Tip Clearance Loss of Impeller Blades", *Journal of Turbomachinery*, Vol 113, October 1991, pp 680-685.

Senoo, Y., 1991, "Mechanics on the Tip Clearance Loss of Impeller Blades", *Journal of Turbomachinery*, Vol 113, pp 680-685.

Williams, R.W., Skelley, S. E., Chen, W-C, and Williams, M., 2001, "Comparison of Unshrouded Impeller Analysis and Experiment", AIAA Paper 2001-3398.

7 FABRICATION AND TESTING OF A SPAR-ACTUATED ACTIVE COMPRESSOR ROTOR BLADE (J. Paduano)

7.1 Introduction and Summary of Results

The active blade was constructed by bonding graphite spars to a high strength-to-weight foam blade. The graphite spars were actuated by piezos bonded on the roots of the spars. The foam blade has the aerodynamic shape of a GE 90 Fan C rotor blade. Strain gauges were attached to the actuators to measure the root strains. Bending mode vibration was obtained by actuating the two graphite spars in phase; twisting mode vibration was obtained by actuating the two graphite spars out of phase. Both modes of vibration can be excited to investigate flutter aerodynamics.

Various technologies were developed to fabricate a geometrically accurate active blade, whose shape duplicates a GE fan blade design. The resulting blade was tested for rotation speed up to 5000 RPM with a vacuum level of 500 micro Torr to 1 milli Torr.

Experiments were conducted before and after the bonding of the foam blade. Bending and twisting deflection results were obtained in four situations **1)** Bench top tests of the graphite spars; **2)** In-situ (spinning) tests of the graphite spars; **3)** Bench top tests of the assembled blade; and **4)** In-situ tests of the assembled blade . The graphite spars and the assembled blade were characterized for their ability to perform broadband excitation experiments.

For the bench top tests of the graphite spars (Bending and Twisting), the results are the following: **a)** For the bending mode, the bandpass bandwidth over which at least 0.5 mm (Peak to Peak) tip bending is maintained 249 Hz, and the bandwidth over which at least 0.25 mm (Peak to Peak) tip bending is maintained 342 Hz. **b)** For the twisting mode, the bandpass bandwidth over which at least 0.5 degree twist is maintained 264 Hz, and the bandwidth over which at least 0.25 degree twist is maintained 354 Hz.

For bench top tests of the assembled blade (Bending, Twisting), the results are the following: **a)** For the bending mode, The bandpass bandwidth over which at least 0.5 mm (Peak to Peak) bending is maintained 38 Hz, and the bandwidth over which at least 0.25 mm (Peak to Peak) bending is maintained 249 Hz. **b)** For the twisting mode, the bandpass bandwidth over which at least 0.5 degree twist is maintained 12 Hz, and the bandpass bandwidth over which at least 0.25 degree twist is maintained 29 Hz.

7.2 Potential Flutter Experiments

In order to investigate flutter instabilities, three potential experiments that test different aspects of flutter are described. These experiments could be conducted with an active rotor.

Experiment I: Investigation of the Effects of Mistuning: Mistuning refers to the blade-to-blade variation of natural frequencies and mode shapes in a rotor, which make it slightly non-axisymmetric structurally. Mistuning may be introduced in a rotor as a means of passive flutter control, and can significantly impact both the stability and forced response of a compressor rotor. The experimental investigation of effects of mistuning parameters is expensive because of the need for different rotors. However, the active rotor can investigate mistuning in two ways. First, geometric mistuning may be studied since altering the angle of twist of individual blades becomes possible. Second, stiffness mistuning can be tested by applying feedback control to the piezos to obtain controllable changes in stiffness of individual blades.

Experiment II: Measurement of Influence Coefficients: Influence coefficients are the basic parameters that quantify the effects of the motion of one blade on another in a rotor. In general, the motion of a blade will lead to pressure perturbations on other blades. The characteristics of these perturbations depend on the mode shape of one blade vibration, the harmonic mode number of the traveling wave of blade vibrations (inter blade phase angle), reduced frequency, and the flow conditions. By using the active rotor, the motion of some blades may be prescribed and controlled, whereas the motions of other blades are passively measured. The generalized forcing function can be calculated based on measured deflections of passive blades. Thus, influence coefficients can be directly measured.

Experiment III: Measurement of the aerodynamic damping: Aerodynamic damping is one of the main parameters used to characterize the tendency toward flutter and severe forced response. It is therefore critical to measure and investigate the effects of the aerodynamic damping. The following steps can measure the aerodynamic damping of a blade: First, the blade dynamics are identified in the presence of flow to measure the total damping, which includes the structural damping and the aerodynamic damping. Second, the blade is identified in a vacuum to determine structural damping. The difference between these damping values is the aerodynamic damping. Currently experimental investigation of aerodynamic damping is expensive. Many different blades are needed to investigate the aerodynamic damping over a wide range of frequencies because each blade can only measure aerodynamic damping at its resonance frequencies. However, the active blade can potentially measure the aerodynamic damping over a much wider range conditions.

7.3 Rotor Blade Construction

Because the piezos do not develop high enough forces to deform a typical titanium compressor blade, a 'spar and shell' concept was chosen. Figure 7.1 shows the 'spar and shell' concept. The active blade consists of two parts: A graphite epoxy core consisting of two spars, and a foam blade (shell). The graphite epoxy core consisting of two spars can survive the centrifugal force. The high strength-to-weight ratio foam blade gives the assembled blade its aerodynamic shape. Figure 7.2 shows how the active rotor uses appropriate phasing of actuation to actively control the shape of the blade.

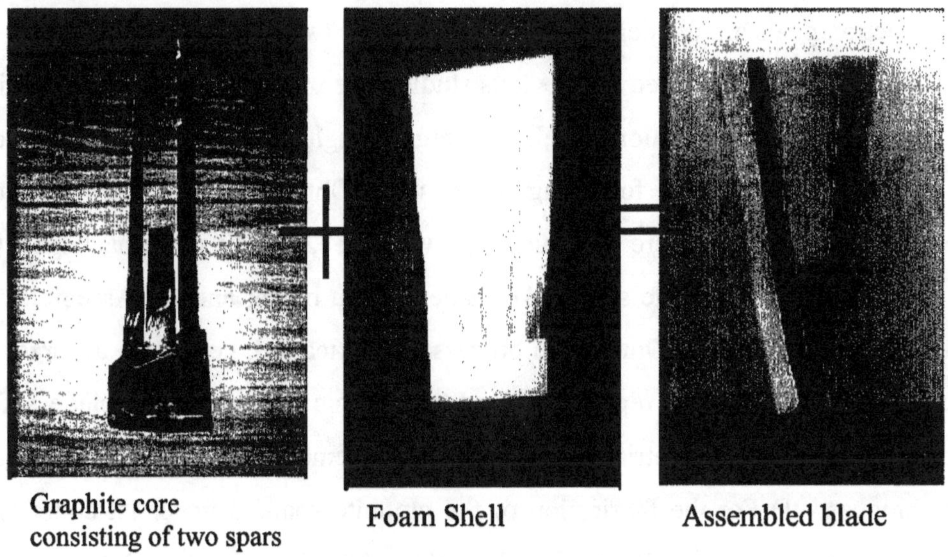

Figure 7.1: Illustration of the 'spar and shell' concept

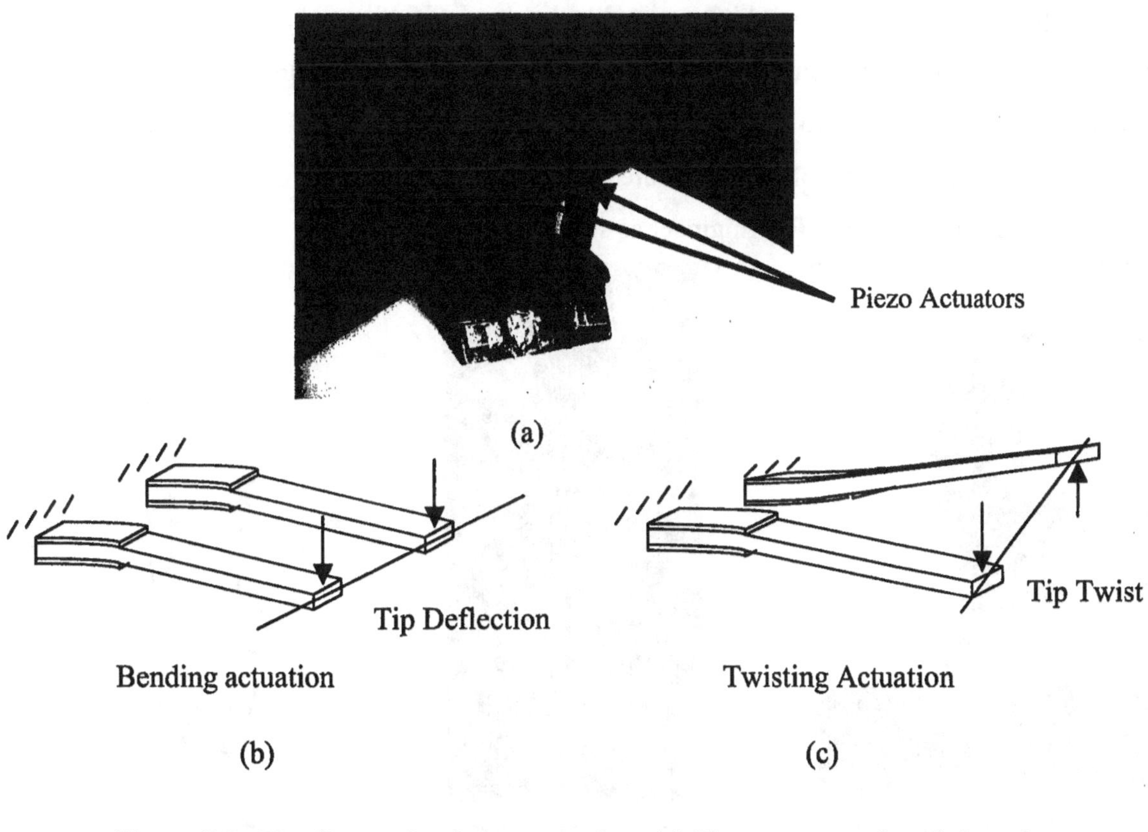

Figure 7.2: Bending and twisting actuation: (a) Piezo actuators bonded on the graphite spars (b) bending actuation (in-phase actuation (c) Twisting actuation (out-of-phase actuation).

The deflection goals are 0.5 degree tip twisting for the twisting mode and 0.5 mm tip bending for the bending mode. Based on previous studies, the induced pressure perturbation around the blade by such deflection would be big enough for system identification. Fabrication can be divided into the following sections: (1) Obtain graphite spars from the Graphite-Epoxy 'Blanks', which were constructed previously; (2) Make piezo packages: (3) Bond piezo packages to the graphite spars; (4) Shape a foam blade; and (5) Assemble the graphite spars to the foam blade. During the process of fabrication, geometric accuracy is critical, since the assembled blade represents the aerodynamic surface of the blade. The following improvements on geometric accuracy were implemented by using appropriate fabrication techniques: 1) For the fabrication of the graphite spars, a water jet cutter was used instead of manual machining. 2) For the fabrication of the foam blade, a 3D numerical milling machine was used instead of thermoforming. Both of these fabrication techniques led to good assembly characteristics and a more accurate geometry of the active rotor blade.

Cutting Spars: The Graphite-Epoxy blank, shown in Figure 7.3, was cut to obtain graphite spars. The blank consists of a complex design and lay up of Hercules' AS4/3501-6 prepreg, molded and cured to meet the geometric and structural specifications of the active rotor. Manufacturing details of this blank are given in previous reports. The geometry of the graphite spars was defined in Pro Engineer as shown in Figure 7.4.

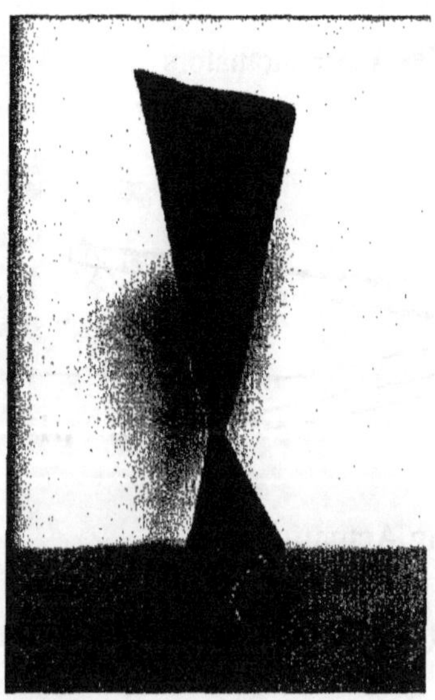

Figure 7.3: Graphite-Epoxy 'Blank'

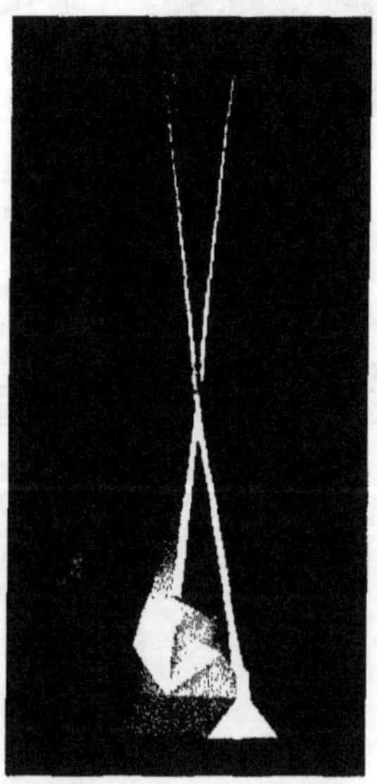

Figure 7.4: Designed graphite spars in Pro Engineer

Figure 7.5: Triangles to be bonded on the Graphite-Epoxy 'Blank'

A water jet cutter was used to cut the spars, using the following new procedure: First the 3D graphite spars geometry was designed in Pro Engineering as shown in Figure 7.4. This design was then projected to a 2D plane to obtain the cutting path shown in Figure 7.6. To support the spars while executing this cut, the blank was attached to a fast prototyped (3D printed) blank mold made by Z Corp as shown in Figure 7.7. Finally, the water-jet cutter was used to follow the numerically defined cutting path to obtain the graphite spars.

Figure 7.6: Obtained accurate projected cutting path

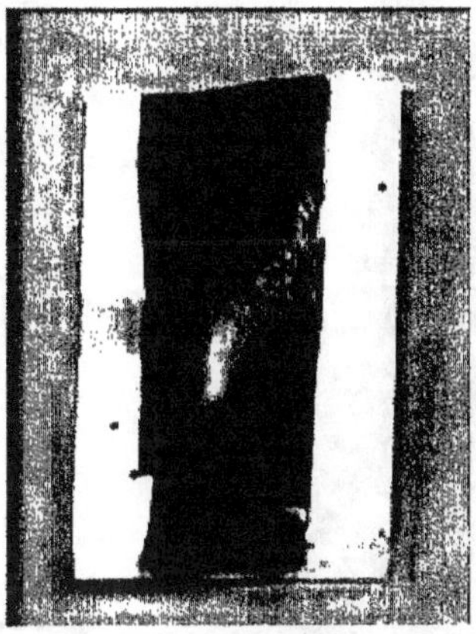

Figure 7.7: Illustration of the attachment of the Graphite-Epoxy 'Blank' to the mold made by Z Corp

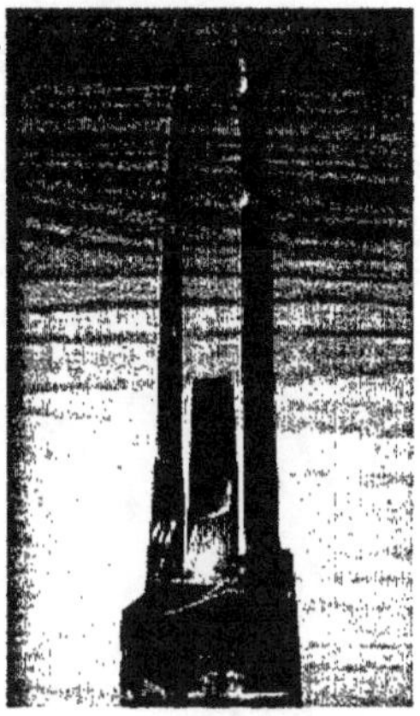

Figure 7.8: Graphite core with cut out spars

The new cutting method yielded very precise graphite spars, which assembled neatly to into the corresponding grooves on the foam blade, which were machined by a 3D numerical milling machine. The cutting method is reliable and quick. In all, three blanks were cut. All the obtained graphite spars are in excellent condition. The machining time for the cutter is less than 1 hour, including adjusting time. Figure 7.8 is a graphite core with cut out spars.

Machining Rohacell: Thermoforming has proved, in the past, unable to obtain the scaled, twisted aerodynamic Fan C shape. Instead, the CNC machine in the Aero Lab at MIT was used to machine a foam block. This method needs a 3-D milling machine, and an aluminum blade mold. The mold is used as a base for holding the machined foam blade when one side of foam blade was completed and the other side of the foam blade was being machined.

Assembling Blade: The previous concept for assembly leads to two piezos embedded inside the composite blade. Figure 7.9 shows that, in this case, the two piezos on the suction side of the graphite spar assembly are exposed after the assembly process. However, the foam blade covered the other two piezos (not shown in the graphite spar assembly picture) on the pressure side after the assembly process.

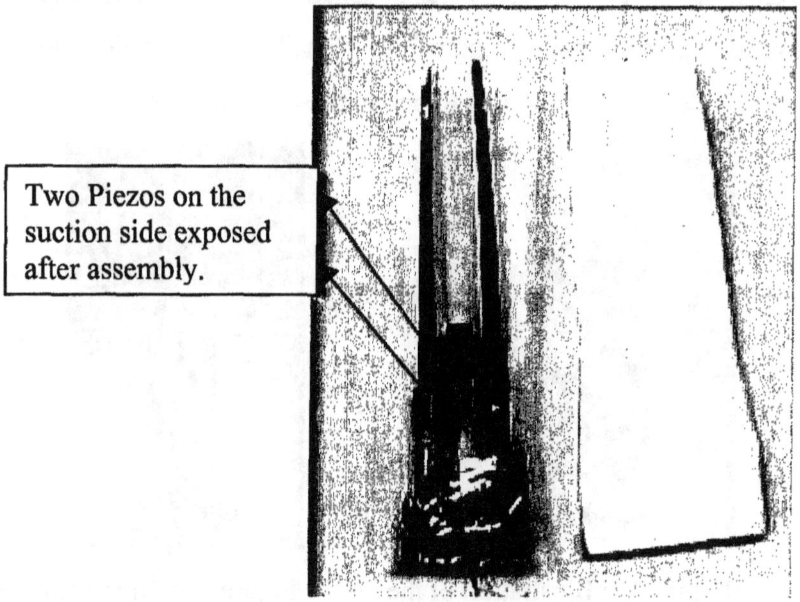

Figure 7.9: Illustration of the piezos to the foam blade

The new idea is to expose all piezos by cutting foam material out at the base of the blade. This approach reduces the risk of breakage during bonding of the foam blade and the graphite spar assembly. It also allows easier maintenance of the piezos, strain gauges, and wiring. Finally, since during the process of bonding the piezos to the graphite spars, more layers of graphite epoxy prepreg were added and then machined to get a flat bonding surface, the geometry of the piezo on the graphite spars has been changed. Hence it was not expected to get good matching of the graphite spar assembly root to the corresponding grooves on the foam blade. Inspection of the assembled blade verifies that the graphite spars and grooves match very well except at the root where the foam material was removed.

The final step was fabrication and bonding of the engineering foam (Rohacell) blade to the graphite spar assembly. Epotec 301 was applied to the cleaned surfaces of the graphite spars and foam grooves. Cure is in room temperature and the minimum cure time is 12 hours. The reason for not using elevated temperature is that the foam blade will stiffen or become soft at temperatures higher than 350°.

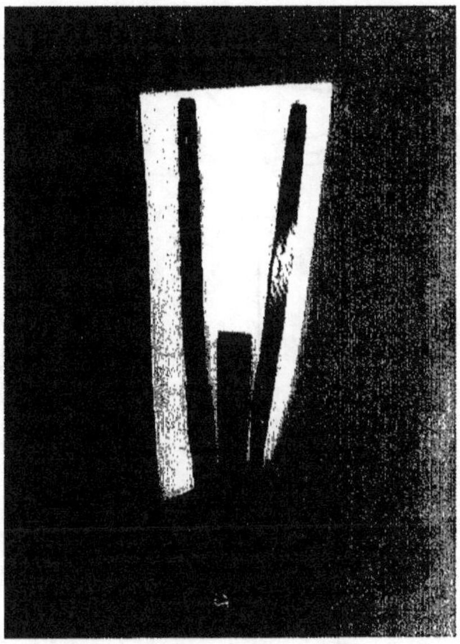

Figure 7.10: Suction side of the assembled blade

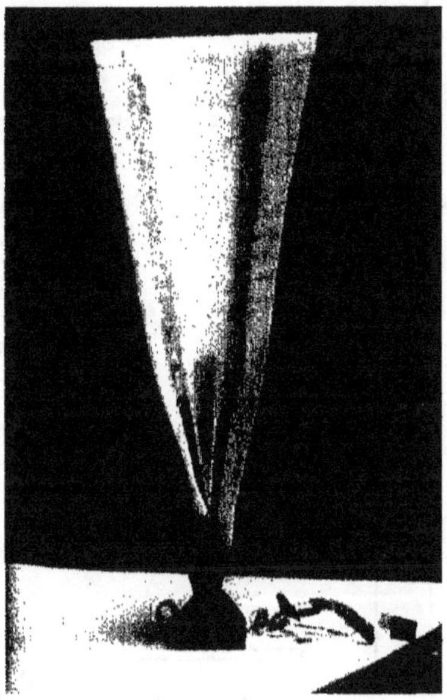

Figure 7.11: Pressure side of the assembled blade

Figures 7.10 and 7.11 show the first geometrically accurate assembled blade. The graphite spars fit well into the grooves on the foam blade. All piezos and strain gauges are working. This is the blade used to perform bench top and in-situ tests.

7.4 Rotor Blade Testing

The instrumentation setup for testing the actuation of the graphite spar assembly is shown in Figure 7.12. The spectrum analyzer outputs a sine sweep, which is connected to both the power amplifier and channel #1 of the spectrum analyzer. This command signal has an amplitude of 1.0 V_{pp}. It is amplified via an audio power amplifier, and its voltage is further raised by transformers. The signal applied to the piezo actuators is about AC 90 V_{rms} with DC 90 V bias. The piezo actuators excite the spars, which in turn deform. Strain gauges measure root strain and a laser displacement sensor measures tip deflections (in bench tests). The output of the strain gauge or the laser sensor was connected to channel #2 of the spectrum analyer. The Analyzer measures the amplitudes of the two input signals, divides them spectrally, and plots the result. Hence, the analyzer can effectively display the frequency response of strain or tip displacement.

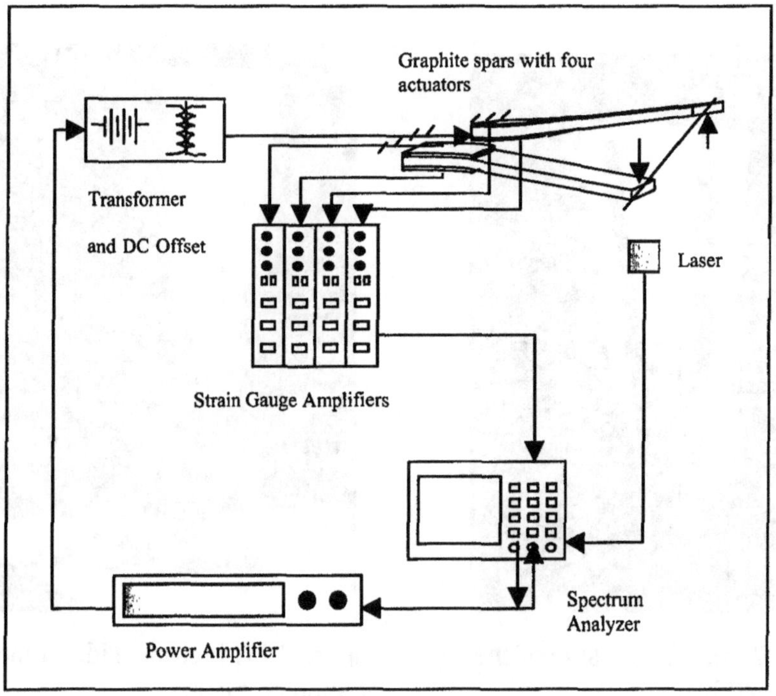

Figure 7-12: Experimental setup for determination of transfer functions

Table 7.1 shows the summary of the measured transfer functions. Table 7.2 shows the inferred transfer functions for the bench top tests of the graphite spar assembly.

Notation:

Strain Gauge at the Leading Edge Spar (LES), Suction Side	s_1
Strain Gauge at the Trailing Edge Spar (TES), Suction Side	s_2
Strain Gauge at the Leading Edge Spar (LES), Pressure Side	s_3
Strain Gauge at the Trailing Edge Spar (TES), Pressure Side	s_4
Tip Deflection at the Leading Edge (LE)	z_1
Tip Deflection at the Trailing Edge (TE)	z_2
Average Tip Deflection (ATD)	$\bar{z}$
Tip Twist (TT)	θ
Command Signal	c

Table 7.1: Summary of measured transfer functions for the bench top tests of the graphite spar assembly

Transfer Function	$\frac{s_1}{c}$	$\frac{s_2}{c}$	$\frac{s_3}{c}$	$\frac{s_4}{c}$	$\frac{z_1}{c}$	$\frac{z_2}{c}$	$\frac{z_1}{s_1}$	$\frac{z_1}{s_3}$	$\frac{z_2}{s_2}$	$\frac{z_2}{s_4}$

Note: In all tests all four piezos are actuated, in phase for bending mode tests and out of phase for twisting mode tests.

Table 7.2: The inferred transfer functions for the bench top tests of the graphite spar assembly

Transfer Functions	$\frac{\bar{z}}{c}$	$\frac{\theta}{c}$

In order to obtain reliable results in the bench top tests, boundary conditions must be similar to the in-situ (spinning) case. So, the graphite spar assembly was mounted on the rotor, which was clamped on the bench. The test frequency ranges from 50 to 1000 Hz. The results of frequency response testing of the graphite spar assembly are presented. Both the leading edge spar and the trailing edge spar were actuated.

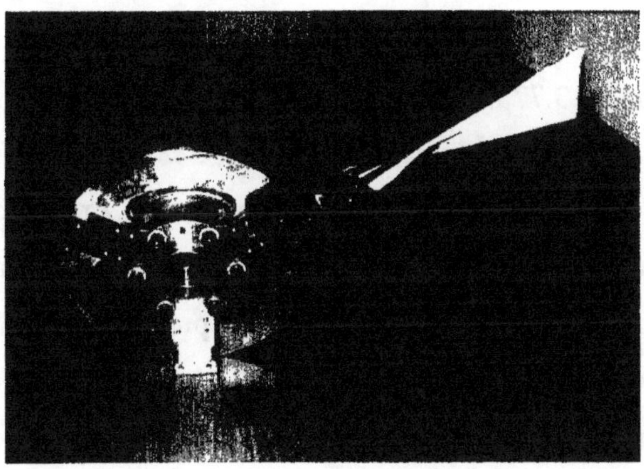

Figure 7.13 – Active Rotor blade mounted in test rotor.

The results can be grouped into four categories. The first sets of results are measured transfer functions from a command signal applied at the piezos to tip deflection at the leading edge spar (LES) and the trailing edge spar (TES). The inferred results of this measured data are the tip bending and the tip twisting transfer functions. The second sets of results are measured transfer functions from the command signal applied at the piezos to strain gauges. The third sets of results are the measured estimation transfer functions, from strain gauges to tip deflection at the leading edge spar (LES) and the trailing edge spar (TES) – these transfer functions were subsequently used in the in-situ tests, where tip displacement could not be measured directly; instead it was estimated using the estimation transfer functions measured in bench tests. The fourth sets of results are the comparisons of the measured and inferred transfer functions from the command signal to tip deflections at the LES and the TES, as well as the comparisons of the measured and inferred tip bending and tip twisting transfer functions.

The measured transfer functions from the command signal to tip deflections of the leading edge spar and the trailing edge spar are shown in Figure 7.14. Based on these measured data, the mean tip defection was obtained by averaging the two spar deflections. The inferred average tip deflection is shown in Figure 7.15. The inferred tip twist was calculated by dividing the sum of the two spar tip deflections by the inter-spar distance at the tip. The inferred tip twist is shown in Figure 7.16.

The following observations about the measured tip deflections can be made:

There are four natural frequencies observed in the frequency range from 50 Hz to 1000 Hz for both the leading edge spar and the trailing edge spar. For the leading edge spar, the natural frequencies are 108.5 Hz, 288 Hz, 576 Hz and 960 Hz with peak-to-peak amplitudes of 16.78 mm/V, 3.77 mm/V, 1.34 mm/V, and 0.6 mm/V respectively. For the trailing edge spar, the natural frequencies are 128 Hz, 288 Hz, 600 Hz, and 900 Hz with peak-to-peak amplitudes of 8 mm, 2.9 mm/V, 1.5 mm/V and 0.2mm/V respectively. We can further conclude that the *bandpass bandwidth,* over which at least 0.5 mm/V peak-to-peak tip bending is maintained 249 Hz, about 92% larger than in previous tests. (See Figure 7.15) The bandpass bandwidth over which at least 0.25 mm/V (Peak to Peak) tip bending is maintained 342 Hz, which is about the same bandwidth as the presented in previous tests. *Bandpass bandwidth* and *bandwidth* are defined in this report as follows:

> *The bandpass bandwidth is defined as the first continuous frequency range over which the response amplitude remains within 3dB of the target value. Other frequency ranges may be found within 3dB of the target value. However, these frequency ranges are not counted into the bandpass bandwidth. Furthermore, note that the frequency response data only start from 50 Hz. No data between the frequency range 0Hz and 50Hz is available. The following criteria are used to infer whether this frequency range is a part of the bandpass bandwidth or not. If the frequency response at 50 Hz is in the bandpass bandwidth, the author infers that the frequency range between 0 Hz and 50 Hz is in the bandpass bandwidth. Otherwise the frequency range is not in the bandpass bandwidth. If the bandpass bandwidth includes the frequency range between 0Hz and 50 Hz, the bandpass bandwidth is called bandwidth. Otherwise it is still called bandpass bandwidth.*

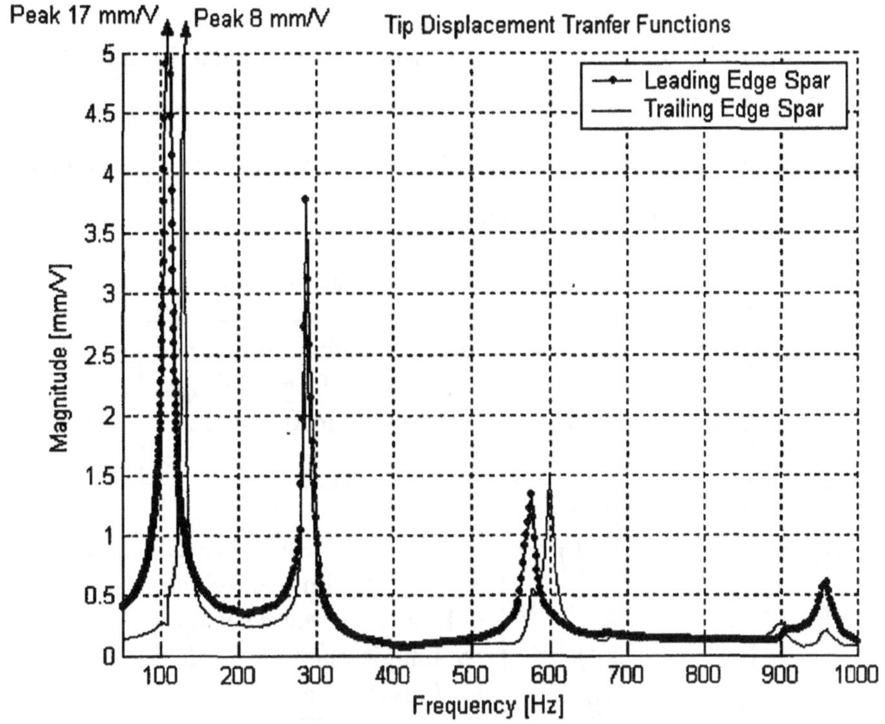

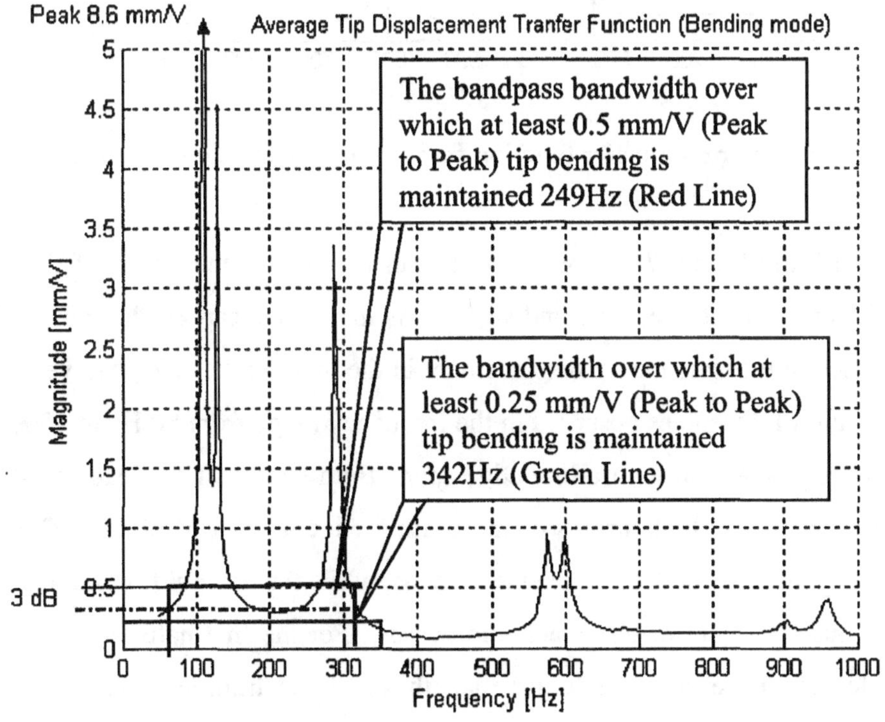

Figure 7.15: Inferred transfer function from the command signal applied at piezos to the average tip displacement (ATD)

Based on this definition, the bandpass bandwidth over which at least 0.5 Degree/V tip twisting is maintained 264 Hz, which is about 210% larger than previous results. (Figure 7.16) The bandwidth over which at least 0.25 Degree/V tip twisting is maintained 354 Hz, which is 136% larger than previously obtained.

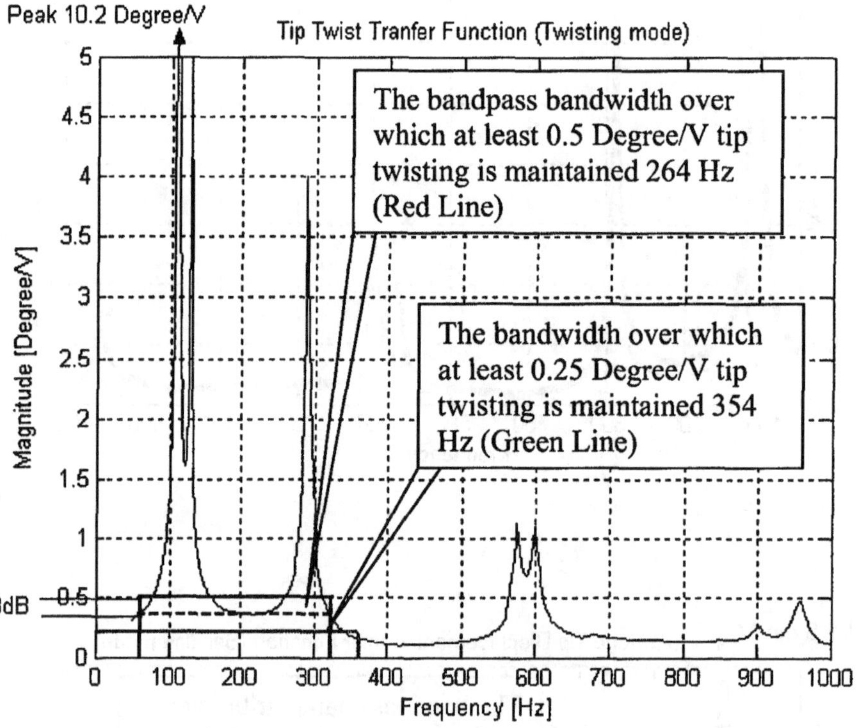

Figure 7.16: Inferred transfer function from the command signal applied at piezos to the tip twist.

Figure 7.17 and Figure 7.18 are plots of both magnitude and phase of the transfer function of the four strains to the command signal. As mentioned earlier, the root strains are the signals we used to estimate the tip displacement in the in-situ test. We observe that there are again four natural frequencies observed in the frequency range from 50 Hz to 1000 Hz for both the leading edge spar and the trailing edge spar. Both strains of the leading edge spar are out of phase. Similarly both strains of the trailing edge are out of phase. This makes sense because when the spar deflects, one side extends, while the other side contracts. This is true for both in-phase and out-of phase actuation cases. For the first natural frequency, the strains of the leading edge spar are larger than those of the trailing edge spar. Correspondingly, the tip deflection of the leading edge spar is larger than the trailing edge spar.

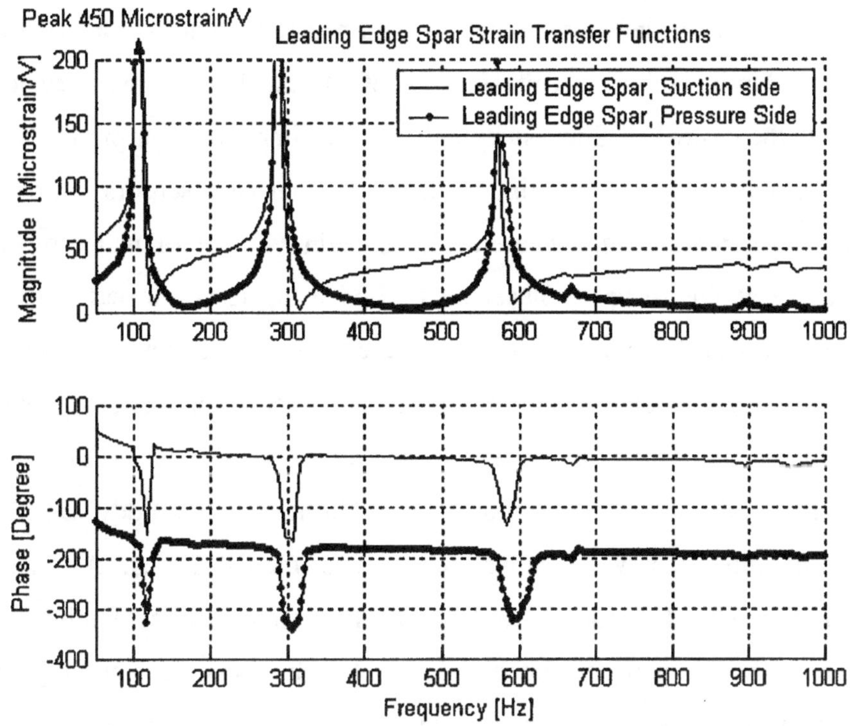

Figure 7.17: Measured transfer functions from the command signal applied at piezos to root strains at the leading edge spar (LES)

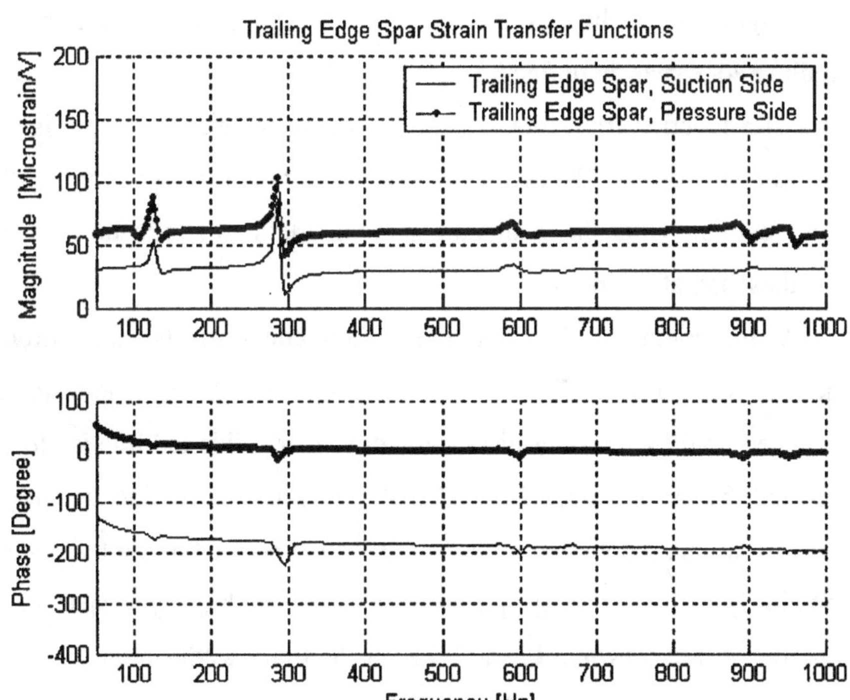

Figure 7.18: Measured transfer functions from the command signal applied at piezos to root strains at the trailing edge spar (TES)

Transfer functions from root strain to tip displacements: Estimation techniques have been developed to determine tip deflection based on strain gauges bonded on the roots of the graphite spars. Until eddy current sensor (ECS) capabilities are developed, and complementary to the desired ECS measurements, such estimation is our preferred method to determine tip displacements in the rotating environment (our laser sensors can only be used in the non-rotating reference frame). These methods begin with determination of the transfer functions from root strains to tip displacements. Bench top tests are performed to obtain these tip estimation transfer functions. Based on these transfer functions, we can estimate the tip displacements in both bench top and rotating tests. In bench top tests, we verify that the results are accurate using laser displacement sensor results. In the RPM tests, only strain-based results are available at the present time. Ultimately the Eddy Current Sensor measurements will help to verify the estimation results of the RPM tests. A series of bench top tests were performed to dynamically characterize the graphite spar assembly.

There are four strain gauges bonded on the root of the spars. Each strain gauge can be used to independently estimate the corresponding tip deflection of the spar. Figure 7.19 shows two estimation transfer functions, which can be used to estimate the leading edge spar tip displacement by measure the root strain of the leading edge spar suction side or pressure side. Similarly, Figure 7.20 showed the other two estimation transfer functions for estimating the trailing edge spar tip deflection.

For the transfer function from root strain to tip displacements, the input is root strain, and the output is tip displacement. Hence, The y-axis is deflection over strain, with units of millimeters per microstrain. Once the strain data is obtained, the tip deflection is very easy to estimate based on these transfer functions.

Figure 7.21 and Figure 7.22 prove the effectiveness of the estimation transfer functions for the graphite spars. In Figure 7.21, the red line is the measured leading edge spar displacement measured by the laser displacement sensor. This red line is plot using the same raw data as that shown in Figure 7.16. The blue line and black line are the estimated leading edge spar displacement obtained by the strain gauge bonded to the leading edge spar pressure side (Strain 1) or the strain gauge bonded to the leading edge spar suction side (Strain 3). Similar results are shown in Figure 7.22 for the trailing edge edge.

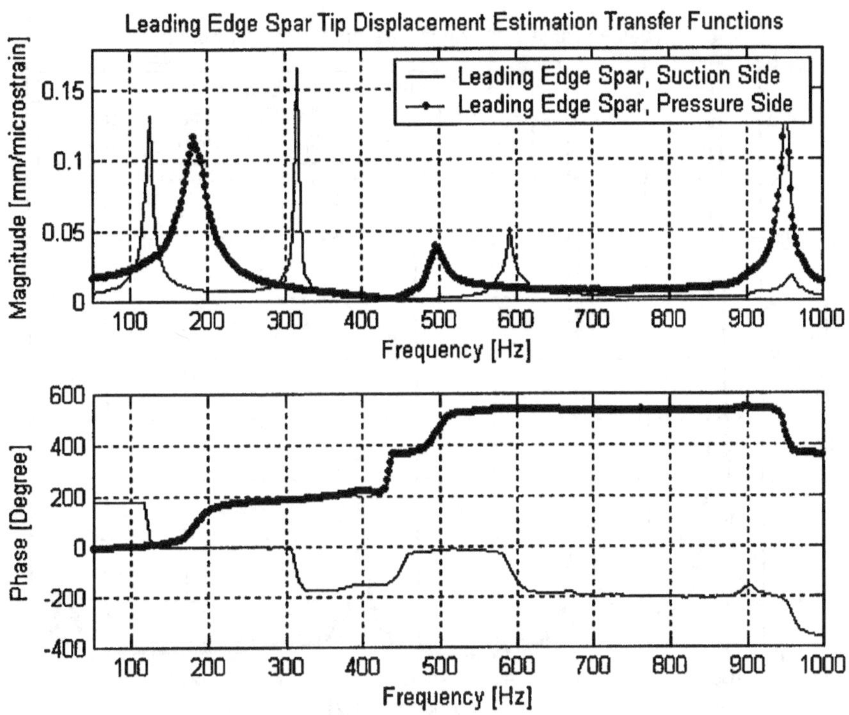

Figure 7.19: Measured transfer functions from root strains at the leading edge spar (LES) to the tip displacement at the leading edge spar (LES)

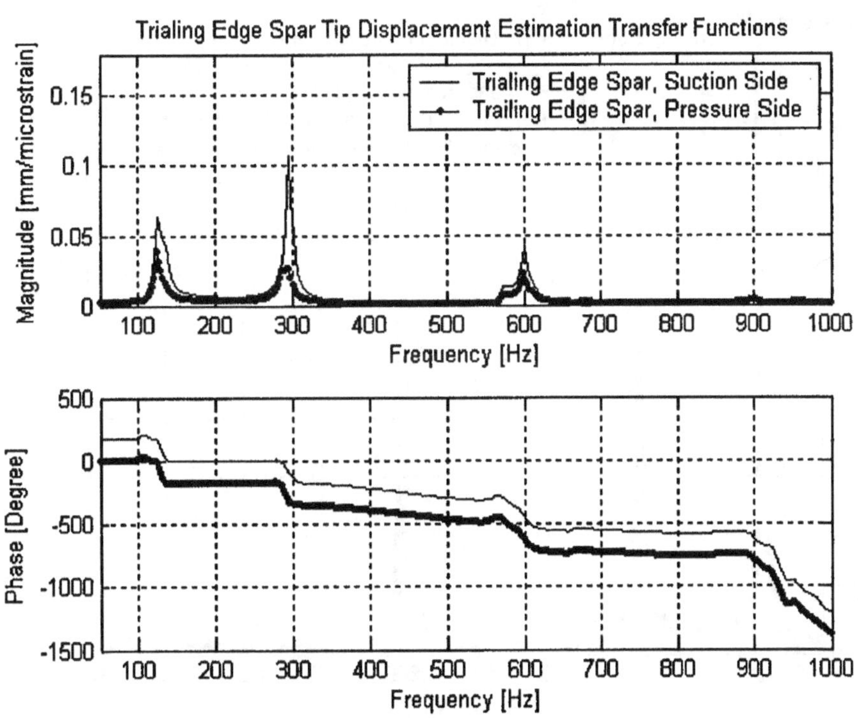

Figure 7.20: Measured transfer function from root strains at the trailing edge spar (TES) to the tip displacement at the trailing edge spar (TES)

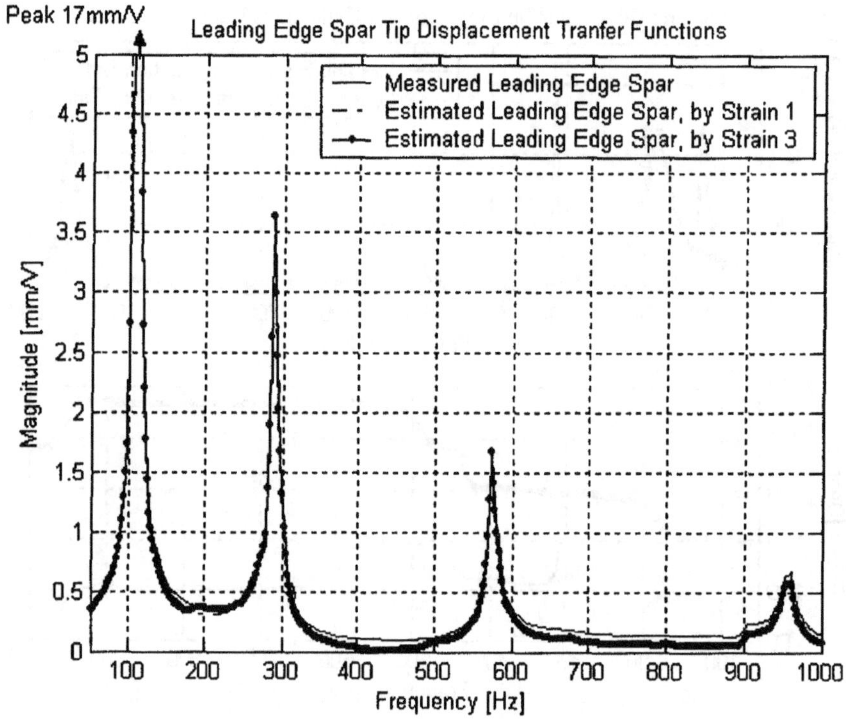

Figure 7.21`: Verification the estimated leading edge spar displacement

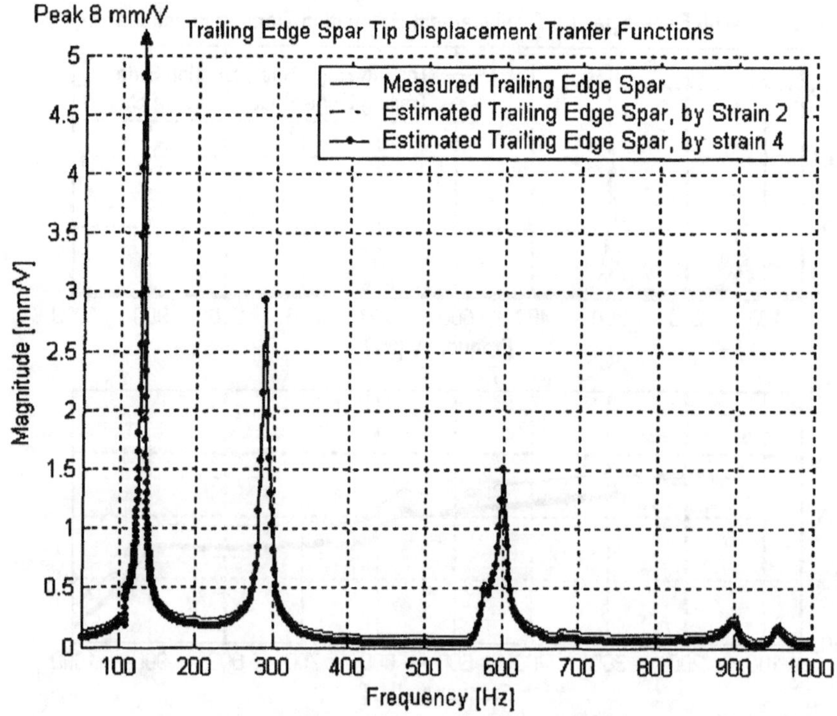

Figure 7.22: Verification the estimated trailing edge spar displacement

Since each curve in Figure 7.21 and 7.22 is the result of a separate experiment, this is a valid test of the procedure. The conclusion is that estimation transfer functions are accurate enough to estimate both leading edge spar and the trailing edge spar tip deflection.

The ultimate parameters of interest for the active rotor are pitch and plunge. Hence it is useful to present the data in terms of the tip deflection and tip twist. Based on the data presented in Figure 7.21 and Figure 7.22, the estimated average tip deflection was obtained by averaging the two spar deflections. The result is plotted in Figure 7.23. The inferred average tip deflection showed in Figure 7.15 is also plotted in Figure 7.23. Figure 7.23 shows that the estimated results using any of the four strain measurements match with the inferred results using the laser displacement sensor very well.

Similar results are shown in Figure 7.24. The estimated tip twist matches well with the inferred tip twist. Figure 7.23 and Figure 7.24 indicate that the state estimation transfer functions are accurate enough to estimate both the average tip displacement and tip twist.

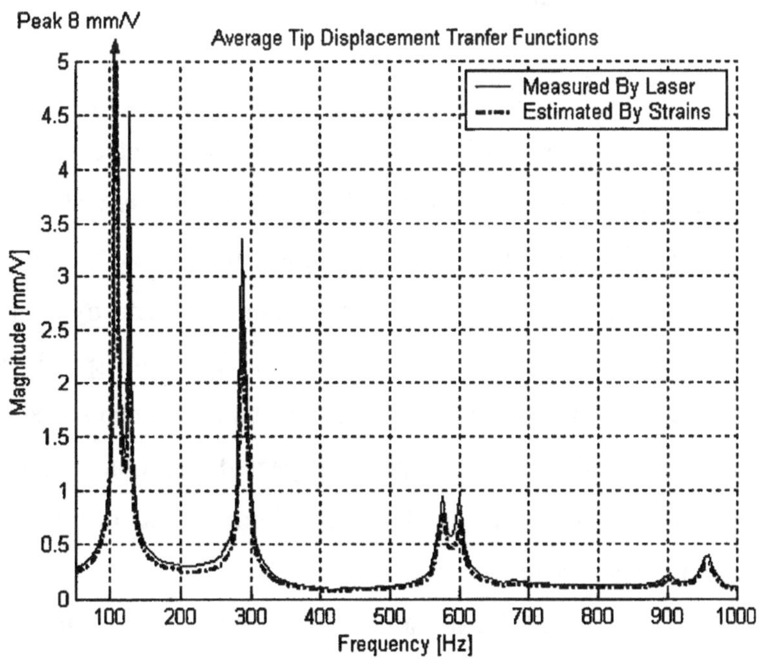

Figure 7.23: Inferred transfer functions from root strains at the leading edge spar (LES) and the trailing edge spar (TES) to the average tip displacement (ATD)

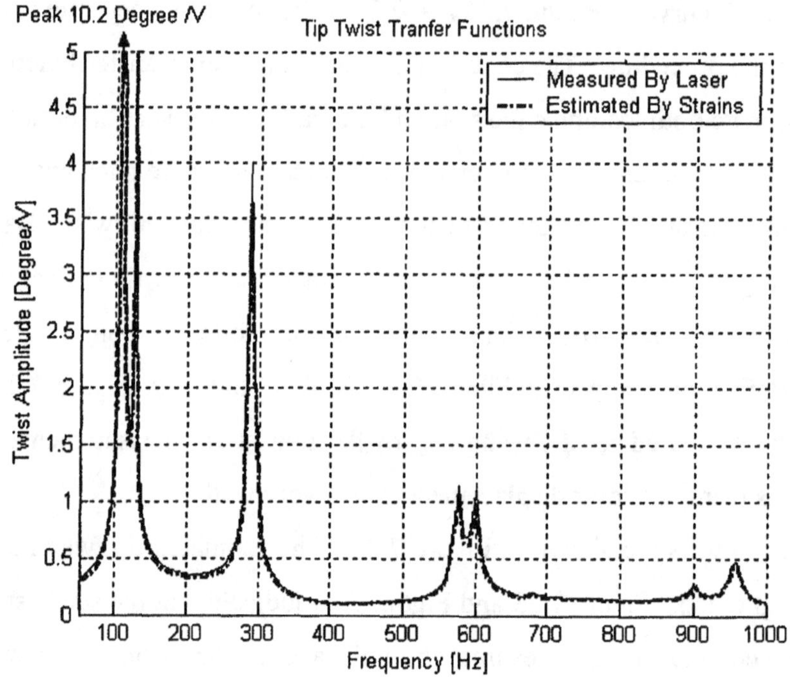

Figure 7.24: Inferred transfer functions from root strains at the leading edge spar (LES) and the trailing edge spar (TS) to the tip twist

<u>In-Situ Characterization</u>: A series of 'RPM tests' were conducted to obtain strain gauge signals from the rotating frame and to determine the effectiveness of spar actuation in the rotating environment. For the RPM tests of the graphite spar assembly, only one graphite spar assembly was mounted on the rotor. Hence both static and dynamic balancing were performed. The rotor speed was first set to 1000 rpm. The rotor speed was then increased in 500 rpm increments, up to 5000 rpm, and data was taken at each speed. When the speed reach 2000 rpm, dynamic balancing was typically repeated, to minimize the in plane one-per-rev vibration. Because of imperfect static balancing there will be a finite rotating imbalance, resulting primarily in a one-per-rev vibration.

Table 3-8 shows the RPM experiments conducted. Since the laser displacement sensor can only be used in the non-rotating reference frame, only four strain gauges can be measured in the spin pit. The average tip displacement and tip twist can only be estimated by using the estimation transfer functions that were obtained in the bench top tests. The inferred transfer functions for the RPM tests are summarized in Table 3-9.

Table 3-8: The Measured root strain transfer functions for the RPM tests of the graphite spar assembly

Transfer Function	$\frac{s_1}{c}$	$\frac{s_2}{c}$	$\frac{s_3}{c}$	$\frac{s_4}{c}$

Table 3-9: The Inferred transfer functions for the RPM tests of the graphite spar assembly

Transfer Funciton	$\frac{z_1}{c}$	$\frac{z_2}{c}$	$\frac{\bar{z}}{c}$	$\frac{\theta}{c}$

In-situ sweep tests, similar to bench experiments, of the graphite spar assembly were conducted. The frequency responses of the graphite spar assembly at different rotating speeds are presented here. The frequency sweep tests yielded transfer functions from the command signal to the strains of the root of graphite spars.

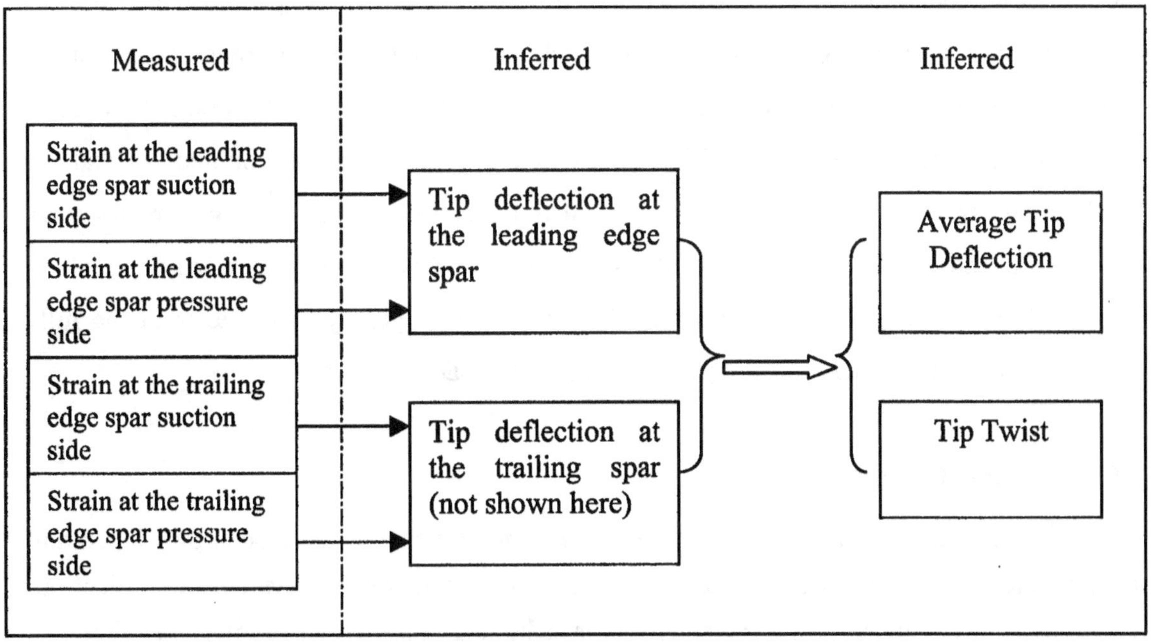

Figure 7.25: A road map for the result of in-situ tests

Since the parameters of interest are pitch and plunge, the raw data of the measured strain gauges are not presented. Only the inferred transfer functions are presented. The excitation frequency ranges from 50 to 1000 Hz. Both the leading edge spar and the trailing edge spar were actuated simultaneously. The following observation about Figure 7.26 are made: For the steady case, one pole dominates the response at a frequency of 108.5Hz, while other less obvious poles are at approximately 288 Hz, 576 Hz, and 960 Hz. These poles experience slight centrifugal stiffening as the rotor is accelerated. By 3000 RPM, the frequencies of the respective poles are 126Hz, 306 Hz, 591 Hz, and 962 Hz. All four transfer functions exibit similarity behavior, so they are not shown here.

Compared with the steady case, the rotating cases have many ripples in the frequency response curves. Background noise is the main reason for these ripples. When the rotor is running, the whole system vibrates. Strain gauges pick up these vibration signals, which are not related to the actuation. The output of the strain gauges is used to calculate the frequency response. Hence these background noise signals appear in the frequency response curves.

Bench-top and in-situ tests of the assembled blade were conducted using exactly the same method as that used for the spars and shown in this report. For brevity, we summarize the results without showing the corresponding plots. First, the methods used are capable of both measuring tip bending and twist on the bench, and using estimation transfer functions to estimate these values in-situ. The stiffness of the foam shell reduces deflections by about a factor two, as one would expect. The modes of the system become coupled, so that the leading edge spar and trailing edge spar natural frequencies combined to form modes at 74 Hz, 216 Hz, and 300 Hz. Spin test results up to 5000 RPM were obtained, verifying that the rotor performs essentially the same up to this operating frequency. Spin tests of the fully assembled blade also verify the structural integrity of the active blade.

7.5 Concluding Remarks

The technologies and processes needed for an active rotor have been developed and demonstrated with attributes needed for aeroelastic diagnostics, system identification, mistuning studies, and control. The active rotor technology provide an enabler, on a far more cost effective basis for: (i) parameterization of rotor mistuning through geometric mistuning (by altering the angle of twist of individual blades) as well as stiffness mistuning (by changing the stiffness of individual blades using feedback control to the piezos); (ii) direct measurement of influence coefficients that quantify the effects of motion of one blade on another in a rotor; and (iii) measuring the aerodynamic damping over a broad range of conditions.

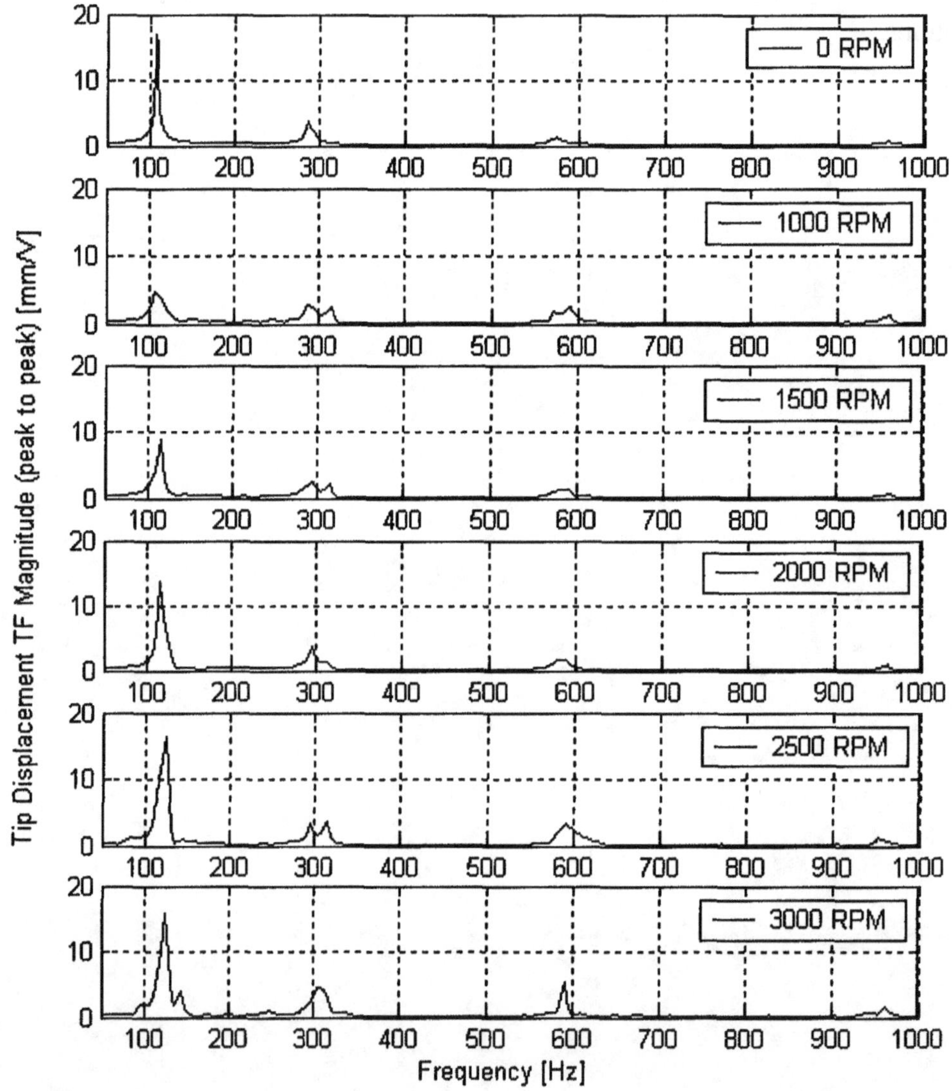

Figure 7.26: Inferred transfer functions from the command signal applied at piezos to the tip deflection at the leading edge spar by the strain at the leading edge spar suction side (RPM Tests)

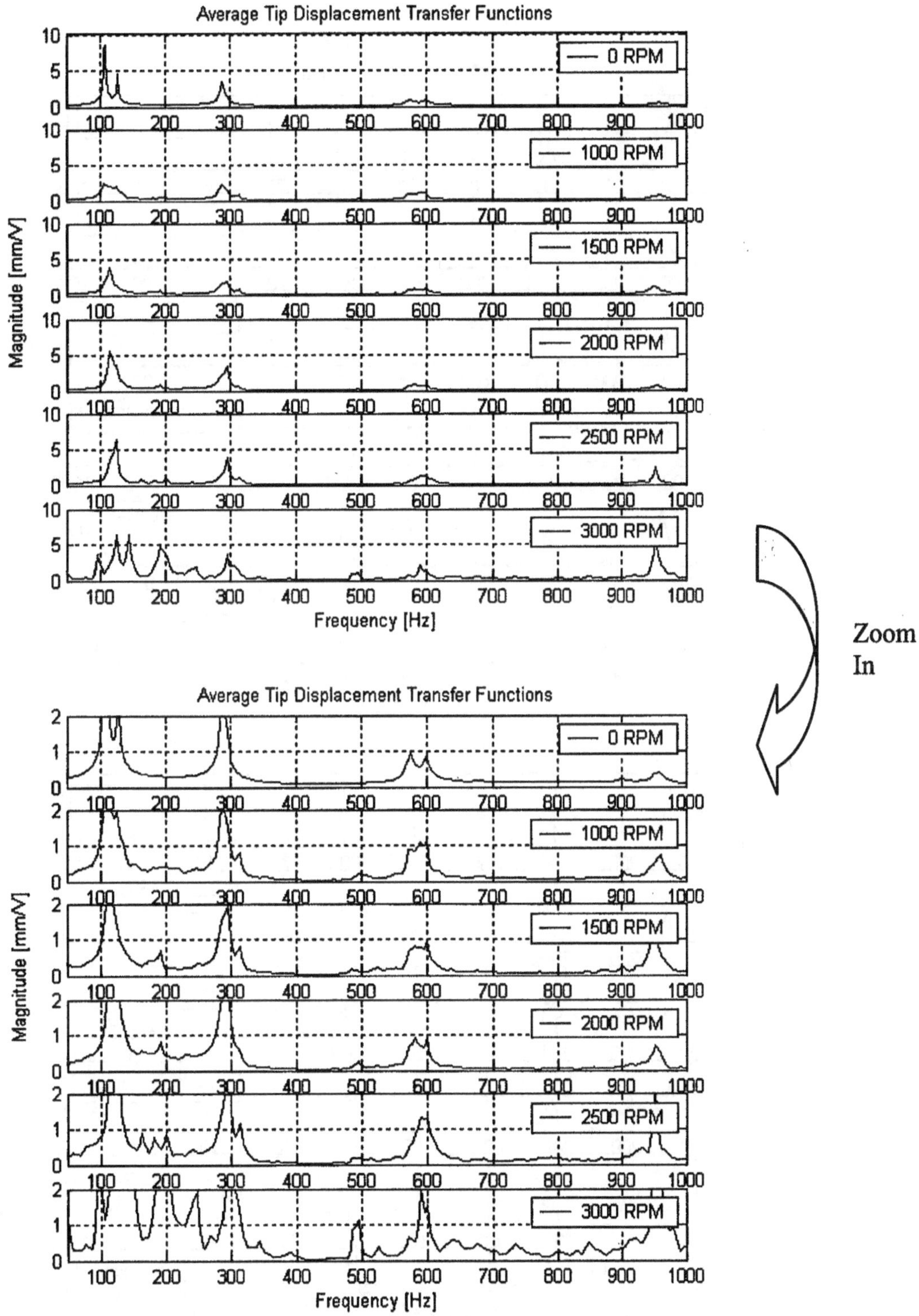

Figure 7.27: Inferred transfer functions from the command signal applied at piezos to the average tip deflection by four strains (RPM Tests)

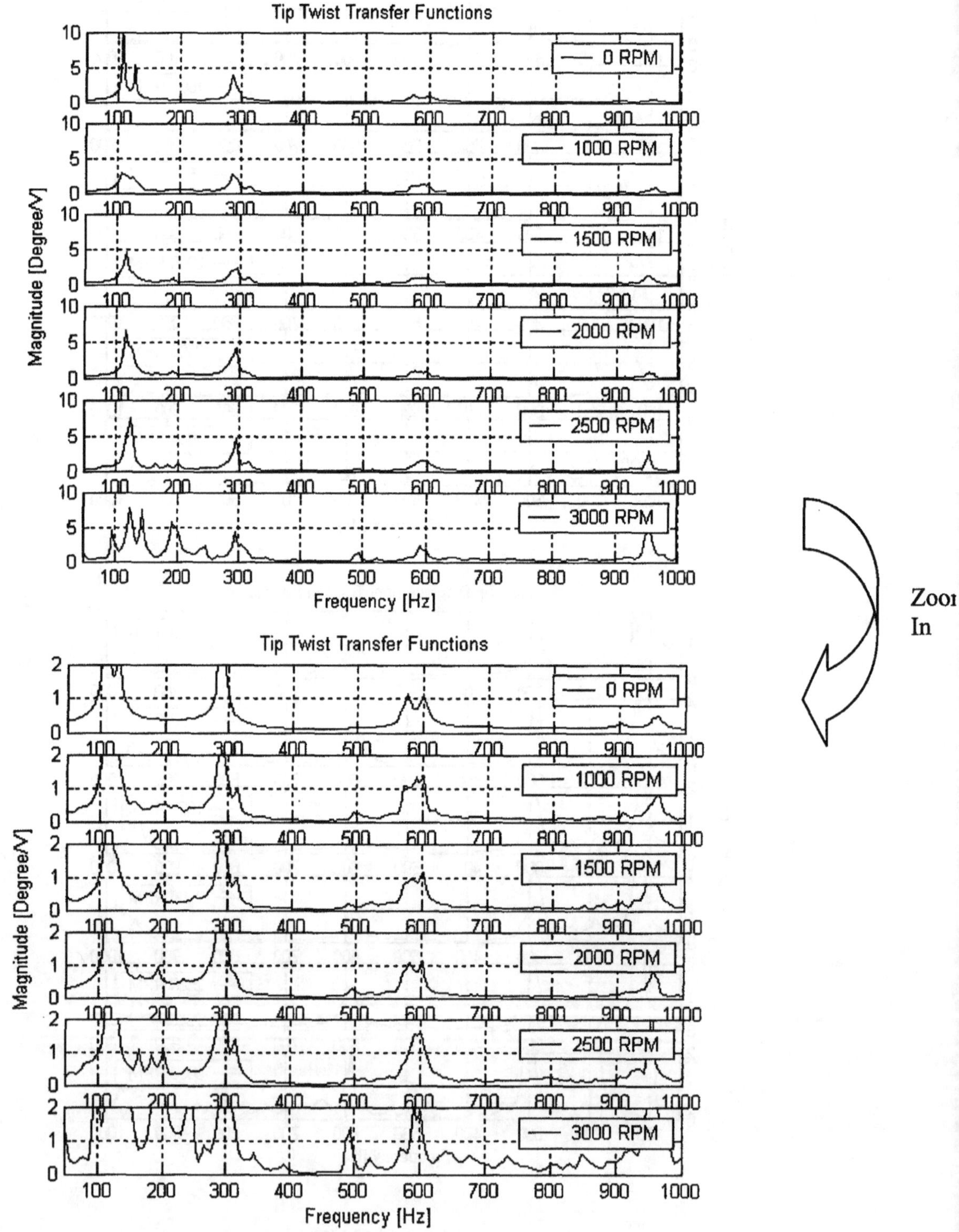

Figure 7.28: Inferred transfer functions from the command signal applied at piezos to the tip twist by four strains (RPM Tests)

8 APPENDIX

Faculty and Staff

Professor A. H. Epstein
Professor E. M. Greitzer
Professor J. L. Kerrebrock
Dr. J. Guenette
Dr. J. Paduano
Dr. C. S. Tan

Graduate Students:

S. Aknouche
A. Khalak
R. Keogh
Y. Lin
H. Vo
Y. Wang

9. PUBLICATIONS

9.1 Doctoral Theses

1. Parametric Dependencies of Aeroengine Flutter for Flutter Clearance Applications by A. Khalak

2. Aerodynamic Performance Measurements of a Film-cooled Turbine Stage by Rory Keogh

3. Role of Tip Clearance Flow on Axial Compressor Stability by H. Vo

9.2 Master Theses

1. Impact of Tip Clearance Flow on Centrifugal Impeller Pump Performance by S. Aknouche

2. Fabrication and Testing of a Spar-actuated Active Compressor Rotor Blade by Y. Lin

9.3 Conference and Journal Publications

1. A Framework For Flutter Clearance Of Aeroengine Blades by A. Khalak
   ASME 2001-GT-270
   Journal of Engineering for Gas Turbine and Power, Vol 124, No. 4, 2002

2. Aerodynamic Performance Measurements of a Film-Cooled Turbine Stage – Experimental Results by R. Keogh, G. Guenette, C. Spadacinni, T. Sommers, S. Florjancic
   ASME GT-2002-30344

9.4 Pending Papers for Conference

1. Role of Tip Clearance Flow on Axial Compressor Stability by H. Vo, C. S. Tan, E.M. Greitzer
   ASME IGTI Turbo Conference 2004

2. Impact of Tip Clearance Flow on Centrifugal Impeller Pump Performance by S. Aknouche, C. S. Tan, E.M. Greitzer
   ASME IGTI Turbo Conference 2004

10. Patents and Inventions

Technology and Process for Manufacturing Articulated Rotor Blades:-

MIT has developed a 'spar and shell' concept for articulated rotor blades. The blade can articulate in bending and twist, at sufficient magnitudes and frequencies for aerodynamic and aeroelastic testing. The concept maintains the 3D blade shape of an industrial transonic compressor, and is structurally capable of withstanding the centrifugal loads associated with transonic tip speeds. By using spars instead of a monolithic construction, articulation is achieved. Specialized layup, water cutting, piezo installation, and shell construction techniques are required to make this blade. Testing up to 5000 RPM was conducted to validate the concept.

www.ingramcontent.com/pod-product-compliance
Lightning Source LLC
LaVergne TN
LVHW082342200926
848481LV00012B/3425